Science and Spirituality
A Hindu Perspective

TO Peter Roe
I am so very happy you attended the party
H. K. K.

H. K. Kesavan

279 Glenridge Drive
Waterloo, Ontario, Canada

September 1, 2003

ISBN: 1-4107-8376-6 (e-book)
ISBN: 1-4107-8377-4 (Paperback)

Library of Congress Control Number: 2003097281

This book is printed on acid-free paper.

Printed in the United States of America
Bloomington, In

1st Books – rev. 09/29/03

Contents

Preface v

Foreword xi

I Basic Concepts 1

1 Some Paradigms of Science 3

2 The Vedas and The Vedic Philosophy 31

2.1 Vedas 31

2.2 The Essence Of Vedic Philosophy 43

3 Some Basic Concepts of Vedic Philosophy 55

3.1 A Conception of Values 55

3.2 Respect for other religions 61

3.3 Śravaṇa, Manana and Nididhyāsana 64

3.4 The Karma Doctrine 66

3.5 Dharma, Artha, Kāma, Mokṣa 69

3.6 Ethics 71

3.7 Concept of Duty 79

3.8 Five Sheaths 83

3.9 AUM (OM) 85

3.10 Īśvara 87
3.11 Meditation (*Dhyāna*) 88
3.12 Māyā 91

4 Science & Vedic Philosophy: Bridging Concepts **95**
4.1 The Fourth State of Consciousness 95
4.2 A Body-Mind Relationship 98
4.3 Some Medical Findings About Consciousness 100
4.4 Several meanings of Reality 102
4.5 Pramāṇa 107
4.6 Evolution 110
4.7 Omega Point 112
4.8 Artificial Intelligence and Consciousness 114
4.9 Some Comments On Cognition And Creation 119
4.10 Space, Time and Causality 120
4.11 Traces Left Behind By The Supreme Truth (Brahmavāsana) 123
4.12 Aesthetics and Spirituality 124

II Metaphysical Theories of Vedic Philosophy **131**

5 Sāṅkhya and Yoga **133**
5.1 Sāṅkhya 133
5.2 Yoga 144

6 Śaṁkara's Nondualism: (Advaita) **153**

7 Theistic Schools: Rāmānuja and Madhva **173**
7.1 A short sketch of Rāmānuja's biography 175
7.2 The metaphysics of Viśiṣṭādvaita 177
7.3 Madhvāchārya: Dualistic School 191

Preface

The topic of Science and Religion has, of late, attracted world-wide attention. For instance, The Templeton Foundation has been very active in arranging workshops on the subject in various universities, inviting participation of scholars from all over the world, and also in rendering financial assistance for the introduction of courses at the university level. I was invited to present a Hindu perspective at a workshop on *The Integration of Heart and Mind* held at the University of Toronto in July 1999. The principal objective of the workshop was to promote healthy discussions between the votaries of science and the broad spectrum of Christian theologians, ranging all the way from the evangelists who strictly adhere to Biblical literalism, to those who did not hesitate to deviate from such rigid orthodox interpretations. Physicists were considered to be the standard bearers for all other scientists, presumably because they were more open to acknowledging the spiritual dimension than, for example, the biologists who subscribe to the theory of evolution due to Darwin as opposed to creationism. The workshop had also a token representation from other faiths, which explained my presence.

In India, there have also been several conferences held on the subject of Science and Religion under various titles: synthesis of science and religion; scientist and the *rishi* (Rishi is a sage or a sage-poet who has attained a high degree of self-realization, or at least, one leading a life completely devoted to the attainment of the spiritual goal); Science and Religion, etc. I had an opportunity to attend one in a series of conferences held under the auspices of the National Institute of Advancement of Science, Bangalore. In the absence of any un-resolvable controversies between Science and Religion from the Hindu perspective, these conferences have been more directed towards determining ways in which scientific education could enrich the understanding of Hindu philosophy.

The book is divided into three parts, the first provides a discussion of some basic paradigms of science and vedic philosophy, as well as the concepts that bridge the two. The second, provide coverage of several major Vedic theories, and the third, provides a discussion on science and spirituality.

Chapter 1 contains a brief discussion of some paradigms of science. We recall some of these ideas at appropriate places in later discussions to serve as a bridge to the understanding of philosophical concepts. Chapter 2 is devoted to a brief introduction to the Vedas and the Vedic philosophy.

Some basic concepts of science and Vedic philosophy are condensed in Chapters 3 and 4. We have also included a section on Art and Spirituality to make these chapters as comprehensive as possible. Each section can be read independently of the others and because of this manner of exposition, there is some inevitable overlap of ideas; however, some repetition is deemed advantageous for the uninitiated reader. The sections are grouped under the headings: a) Vedic Philosophy; and b)Science and Vedic Philosophy: Some Bridging Concepts. Even those who are well versed in Indian philosophy have found these materials illuminating, and they can be read with a basic understanding of the previous chapters.

Chapters 5, 6 and 7 deal with five metaphysical theories within the Vedic fold and are intended to provide an overall understanding of the vast subject. Sāṅkhya and Yoga are taken together in Chapter 5 because of their kindred relationship. Chapter 6 is on non-dualism due to S̃aṁkara, which is sometimes referred to as an absolutistic model for realizing the supreme reality, and Chapter 7 is on theistic models due to Rāmānuja and Madhva in the tradition of devotion and abject surrender to the Lord. Since these chapters appeal only to those who are looking for an in-depth understanding of Vedic philosophy, it is advisable to skip them at least on the first reading of the text. The novelty of the treatment, however, is to make these treatises accessible to those who have not undergone classical education characteristic of traditional studies on the Vedas.

Chapter 8 is a free-wheeling discussion on science and spirituality and points to the need for a greater understanding between different faiths based on some universal concepts underlying *sanātanadharma* without even a remote suggestion of the necessity for converting from one faith to another. We expect the majority of the readers to be interested in reading Chapters 1, 2, 3, 4 and 8 at least on the first round.

The book is concerned with highlighting spirituality that underlies Vedic philosophy in order to portray its Universality; discussion of Vedic religion is outside the scope of this book. It is my fond hope that this special approach will serve to illustrate the manner in which Vedic philosophy resolves the seeming conflict between science and religion, as well as provide a basis for promoting a healthy inter-faith dialogue from the common platform of spirituality.

Although my target reader is one who is generally interested in the broad topic of science, philosophy and spirituality from the Vedic perspective, I also have a special audience in mind because of my own background as a person of Indian origin living in North America for more than four decades. Based on my contacts with people of similar background, I believe this text would be of special interest to them. What I say next in support of this observation might appear as a digression, but, hopefully, it will not detract from the broader appeal of the book.

Indians living abroad have to deal not only with routine matters like professional opportunities and financial well-being but must also wrestle with the deep-seated concerns that arise from their anxiety to retain at least a significant part of their Indianness, which they obviously relish with a great deal of nostalgia. Moreover, they know in their heart of hearts that such an identity is absolutely essential for remaining a recognizable part of the cultural mosaic of these western countries. The problem, therefore, is to find skillful ways of maintaining the essential aspects of their immigrant culture, while at the same time to remain part of the social fabric of the countries of their adaptation.

While living in India, many of the first generation immigrants developed an appreciation of their distinctive culture, not through formal instruction, but through osmosis. They were seldom challenged to justify their views on important matters, thus never having to bestow serious thought upon matters relating to their heritage. As immigrants, however, they found themselves in the rather unenviable position of having to explain the rationale for their patterns of thought and modes of behavior to their own children, who are brought up in these countries, as well as to the society at large which cannot be expected to have an understanding of their religious and cultural history. This difficulty severely strains the majority of the immigrants, who either are not used to intellectual inquiry or do not have the time and patience for what they consider esoteric matters. Children, however much they might want to explain away their parents' shortcomings, are the first ones to observe that the older generation are in no position to provide satisfactory answers to questions, raised in all earnestness, about Indian heritage.

Living in societies with an entirely different cultural ambience, parents cannot afford to lament over the lapse of the authoritarian model which governs, or at least used to govern until lately, the Indian family structure, which had once shielded them from such embarrassment. Within more contemporary family setting, they cannot easily brush aside the questions their children pose, particularly since the parents ardently wish their children to inherit the essential values of their own rich heritage they have brought with them. This inherent conflict places them on the horns of a dilemma. On the one hand, they want to make sure that the younger generation imbibe their cultural values, but on the other, most of them do not have a clear idea about what they want their children to inherit. While their drive to better their educational and financial prospects, which brought them to these countries in the first place, is still alive, nevertheless, it is eclipsed by their desire to seek a deeper meaning to their lives and indeed to those of their posterity. Thus, living a completely integrated life becomes their chief concern.

On introspection, we find that of all the factors that have shaped our cultural identity, our Hindu religion is the major determinant in establishing our special way of life. Untutored as we are about the essential teachings of the religion, we doggedly cling to the traditional practices of the faith with little or no appeal to reason, as if they provide the only glue to the society that we have left behind. Very soon, delusion sets in due to the realization that such practices cannot easily be transplanted to these countries because of differences in environment and, more so, because of a lack of enthusiasm from the younger generation. On deeper inquiry, we start to look for those fundamental aspects of our culture which are invariant with respect to environmental conditions so that they have a real chance of taking root in our new habitats.

The realization gradually sets in that the Hindu religion cannot be successfully practiced without a proper understanding of its underlying philosophy, particularly outside of India because of the absence of the weight of its tradition and the scarcity of knowledgeable people who can serve as guides. However, pursuing the analysis based on our new experience, we suddenly see some light for the resolution of our dilemma because of the conviction that if Hindu philosophy has an intrinsic universal appeal, as it professes to have, then it is the element of spirituality that provides the much sought-after portability of values. My discussions with the university students of Indian origin at the University of Waterloo were along these lines, and the questions they raised greatly helped me to clarify my own thinking on science and religion and many allied topics. In particular, I found that the paradigms of science were extremely useful in making contact with students with a good university education. This conceptual framework is reflected in the development of this book, which is perhaps its novelty.

I have a long list of people to acknowledge for helping me develop my own thinking on the subject of Vedic philosophy right from my student days. My initial interest in the fascinating subject was kindled by my professor D.S.Subbaramaiya, who taught mathematical physics with distinction in Central College, which is located in my home town of Bangalore. He was a research student of Dr. C.V.Raman, a Nobel Laureate in physics, at the Indian Institute of Science, Bangalore, before entering the teaching profession. He later on became a full-time devotee of the S̃aṁkara tradition and was especially requested by His Holiness, the Jagadguru of the Sringeri monastery, to write a commentary in English on the Dakshinamurthystotram, which is a superb composition of Adi S̃aṁkara. This scholarly book, written by an ardent practitioner of the faith, has appeared in two volumes which are included in the list of references. His background in mathematical physics is very evident in the commentary. I had the privilege of maintaining contact with him throughout the years till his demise a few years ago.

My next acknowledgment is to the *Ashtanga-yoga-mandiram* in Bangalore. Its original preceptor (who is now no more) is Srirangaguru, who is regarded by many of his disciples as not only a man who attained self-realization, but also as someone who successfully led others to attain a higher level of consciousness through his remarkable spiritual abilities and infinite compassion. Although he had very little formal education, his scriptural knowledge was encyclopedic, and he had the matching talent for clear articulation in the Kannada language. He never asked for a dogmatic acceptance of his teachings, but instead emphasized the need for understanding them on the basis of experience. His divine spouse, who has also attained an exalted level of spiritual perfection, is now the titular head of the ashram. All the active work of the establishment is now done by the great master, Shri Rangapriya Swamiji, who, before becoming a renunciant, was a professor of Sanskrit at the National College, Bangalore. It was my good fortune to have come in contact with him twenty years ago.

At Waterloo, there is a Brahmarishi Mission which was established twenty years ago by Swami Brahmarishi Visvatma Bawra, with headquarters in Haryana, close to Chandigarh. The Swamiji used to visit Waterloo every year as part of his itinerary that included visits to his other centers in Canada, the US, UK, and some other countries till his demise three years ago. He was a great expositor of Sāṅkhya philosophy, and it was difficult not to be influenced by his remarkable presence. He also taught meditation as a matter of course just prior to the commencement of his captivating lectures delivered in his native Hindi language with great flourish and embellishment. Needless to say, I have greatly benefited from my contact with him.

I do not profess to be an expert on Indian philosophy because my principal occupation has been that of a professor in the engineering faculty. Whatever I know with some degree of coherence, apart from what I have learnt from my numerous encounters with knowledgeable people, I owe chiefly to the books written by the late professor M. Hiriyanna, which I have cited in my list of references. I discovered very recently that he was very highly regarded for the depth and breadth of his scholarship and also for the precision and clarity with which he spoke and wrote. He was a colleague of Dr. S.Radhakrishnan, the more popularly known Indian philosopher and later president of India, in the Maharaja's College, Mysore. Since he belonged to an earlier generation, I never had the opportunity to meet him in person.

The list of people from whom I have gained an understanding of the broader subjects of science, philosophy and religion is too long to mention. In my own teaching of system theory at the University of Waterloo, I have also been greatly influenced by several Western philosophers. Most certainly, the scientists that I have listed in my reference pages and particularly the physicist and popular science writer Paul Davies, and the Princeton physicist Freeman Dyson, have influenced my decisions on the topics of Chapter 1. The challenge was to select a short list of scientific paradigms from a vast array of available ones which had direct relevance to the later development of the Vedic philosophy. The major milestones in scientific thought have undoubtedly resulted in pronounced shifts in our understanding the external universe; consequently, they provide deep insights into the study of the Vedic philosophy.

My sincere thanks are due to New Age International, New Delhi, which published an earlier Indian edition of this book under the title, Science and Mysticism: The Essence of Vedic Philosophy. I also wish to thank my daughter Kalpana Sarathy and her husband Dr. Sriprakash Sarathy for their invaluable assistance in getting this greatly revised book published. I have also benefited from published critiques of my earlier edition and from the valuable comments I have received from several readers.

Waterloo, Ontario, Canada
September 1, 2003

Foreword

The Hindu viewpoint on Science and Religion is unequivocal: it does not posit any inherent conflict between the two pursuits. The Sanskrit word *vijṇāna*, which is of ancient origin, is commonly translated as science, a word of relatively recent origin. *Vijṇāna* is closely related to its complementary word, *jṇāna* (God consciousness), which is the final goal of religion. The domain of *vijṇāna* is the rich diversity of creation, whereas the domain of *jṇāna* is the unity behind the diversity. The ultimate purpose of *vijṇāna* is to take one towards *jṇāna* and so, from this standpoint, *vijṇāna* has a much broader meaning than what is implied by the word science. In fact, the division of secular knowledge into the various branches of *vijṇāna* that appear in the classical Hindu literature is based on their relationship to *jṇāna*.

The steady progress of science in search of the ultimates has resulted in major shifts in our understanding of the external universe. While these paradigms of science have, undoubtedly, enriched the development of Western philosophy, they have had negligible influence on Hindu philosophy because of their entirely different origins. We proceed to introduce some background material to substantiate this assertion.

Hindus themselves would prefer to call Hinduism *sanātana dharma,* translated as eternal religion. This is important to note since Hindus do not have an organized church, a single prophet, or a single holy book; the Vedas which are the Hindu scriptures, constitute a whole literature covering a wide spectrum of ideas ranging from the secular to the spiritual. They are at least older than the time of Buddha, who died in 487 BC, a firm date in Indian history. Unfortunately, we can only set the minimum and maximum bounds derived from inferential knowledge, because of lack of proper historical records. Historians place the period of the Vedic age anywhere between 1500 to 2000 BC. The timeless message of *sanātana dharma,* which is about life comprising both secular and spiritual values, is addressed to every individual. Hinduism does not claim its religion to be either unique or exclusive; no single prophet or single holy book is involved either in the exposition of its theoretical knowledge of the ultimates, or in the practical means laid out for its achievement. The theoretical knowledge is called the Vedic philosophy, and the practical aspects are embodied in the Vedic religion. Consequently, Vedic philosophy has a different status than that of Western philosophy whose role, according to Bertrand Russell, is to fill the grey area between theology and religion.

Unlike Western philosophy, the fundamental tenets of Vedic philosophy were not in any way, influenced by the developments in Natural philosophy. This does not mean that interest was lacking in unravelling the mysteries of nature, particularly those concerned with the origin of life, origin of consciousness, origin of the universe, etc. We single out these topics since they have engendered relentless controversy between science and religion in Christianity in particular as evidenced by the endless debate on creationism and biological evolution. However, the major milestones in the advancement of science were not viewed as the essential steps in the development of Vedic philosophy. The latter adheres to this viewpoint, because philosophical inquiry is about the ultimates —a study of reality as a whole, unlike scientific inquiry, which is interested in fragments of reality. Exclusive attention is paid to the study of nature (the legitimate domain of science), however, it has the unfortunate effect of obscuring the importance of a holistic viewpoint (the concern of philosophy),

In Western philosophy, Rene Descartes is given the credit for highlighting the dichotomy between *mind* and *matter*, between the subject *I* and the *rest of the universe*, between *being* and *becoming*. These ideas are also firmly embedded in the development of Vedic philosophy. For instance, Sāṅkhya philosophy, one of the six branches of Vedic philosophy which pre-dates Buddhist times is an elaborate and complete thesis on spirit, matter and their interrelationship. We have devoted an entire chapter to this philosophy because it lays out most of the fundamental theoretical concepts of all branches of Vedic inquiry. The appreciation of the concept of duality between spirit and nature is a penultimate step in the understanding of the concept of unity that runs throughout the discussion of later developments of Vedic philosophy. The notion of *I* , which Man experiences throughout his lifetime, is the most stubborn thought that he encounters; the empirical I, the human ego, is referred to as *ahamkāra* in Sanskrit.

With the above background, we can state further reasons why Vedic philosophy does not lean on the understanding of natural philosophy. This non-dependence on natural philosophy constitutes an important point of bifurcation from Western philosophy. One could start investigating the element of constancy associated with the external universe by studying the natural laws and attempt to arrive at a theory of everything (TOE). Alternately, one could investigate the true nature of I, by delving into what constitutes the element of constancy in the feeling of I in all the states of consciousness, namely, the waking state, dream state and the deep sleep state. Or, as it is stated in cryptic fashion, "*who am I*" becomes the central question to be answered. Vedic philosophy has deliberately chosen the second line of investigation, by setting aside interest in natural philosophy as a positive aid for this specific inquiry. This deliberate choice also qualifies as a holistic inquiry since its final conclusion obviates the need for a Cartesian dichotomy between *mind* and *matter*. The Vedic inquiry into the true nature of I should not be misconstrued as a study in psychology, since all that we are interested in are some universal features of the mutually exclusive triad of states of consciousness in the study of the human condition as a whole. Though it is related to our psycho-physical apparatus, the study is best described as the science of the soul. In India, philosophical speculation has claimed a parallelism between the search for ultimate truth behind the universe as a whole and the search for the ultimate truth behind the empirical ego of man. The details of this parallelism make for a very fascinating study. It is central to the study of the *Upanishaḍs*, which constitute the final sections of the Vedas.

Scientific inquiry, as well as all other secular studies, are categorized as *aparā vidya* (lower truth), the study of the phenomenal world, whereas the search for the ultimate truth behind the feeling of I, or the truth of the transcendental realm, is categorized as *parā vidya* (higher truth). There is no order of importance suggested in this classification; in fact, the two are considered to complement each other as knowledge itself is considered to be one continuum.

Vedic philosophy suggests why scientific inquiry runs into an impasse in its attempts to unveil the ultimate truth behind nature. First, a scientific investigation is not about the whole of reality; the laws of nature, however grandiose in scope, merely provide an intellectual understanding of some specific aspects of nature. Furthermore, the sum of all scientific truths about the external universe does not amount to the inquiry about the whole truth since it does not probe into the true meaning of I, which belongs to the non-manifest field of existence. Secondly, a careful observation will reveal that, as a matter of methodology, one cannot *a priori* invest the whole truth to nature and then start investigating it by relying on valid means of acquiring knowledge (*pramāṇas*), such as perception, inference etc., which are themselves ingredients of nature. Many logical paradoxes are known to start from similar false premises. It is as hopeless as trying to stand on one's own shoulders. Because of these considerations, Vedic philosophy, while it recognizes the importance of the *pramāṇas* for acquiring scientific knowledge, declares the need for another *pramāṇa* over and above those of the earlier types to plumb into the transcendental realm. In the case of the *sanātana dharma*, this additional *pramāṇa* is the Vedas. The relationship of a valid means of acquiring knowledge in the scientific realm to its counterpart in the transcendental realm, is carefully laid out without giving rise to logical fallacies. For instance, it is illogical to invoke the testimony of Vedas to deny or validate truths that belong to the scientific realm. This is why, for example, the raging controversy between creationism and evolution does not arise in the Hindu perspective.

Implied in the assertion that the Vedas constitute a *pramāṇa* for gaining knowledge of the non-manifest field of existence is the important conclusion that, in the ultimate analysis, it is only spiritual knowledge that can remove our ignorance about the ultimate reality and not merely mystical experience, as is often, mistakenly, claimed. That the sun rises in the east and sets in the west is an experience that appeals to our common sense, but such an experience has to be assimilated within the framework of knowledge of planetary motions. Similar caution has to be exercised when interpreting mystical experiences, important as they are in the progress towards spiritual perfection.

It is because Vedic philosophy seemed to be completely divorced from natural philosophy in the sense discussed above, that many American schools, even those devoted to inter-religious studies, had difficulty accepting its inclusion in their curricula as a legitimate field of study. But now that the spiritual iron curtain that was drawn in ancient Greece (to use one of Aldous Huxley's picturesque phrases), has been lifted, there is increasing interest evinced for the study of Eastern philosophies in general. Buddhism, the sister religion of Hinduism, was accepted first, and gradual acceptance for the study for Hinduism followed later. There are cynics who suggest that scientists were drawn into the study of Buddhism, because it does not deliberately invoke the presence of a supreme being, and as for Hinduism, its attraction is because its central theme is about the existence of one supreme reality. But, cynicism apart, the study of science and religion in its totality will not be complete if the religions indigenous to India, are left out of the equation.

The question can be legitimately raised why scientific studies are germane at all to the understanding of Vedic philosophy and Vedic religion. The first answer is that Vedic philosophy does not deny the importance of worldly realities and therefore has absolutely no intention of impoverishing itself by denying the role of science in its legitimate realm. Furthermore, strange as it may seem, the study of science also offers profound insights into the understanding of the intellectual origins of Vedic philosophy. Although the transcendental realm, the realm of God consciousness, is even beyond the limits of rational thought, science, however, does provide sharp pointers to the understanding of spiritual knowledge. By this we do not mean that we seek spurious analogies from science for understanding truths of Vedic philosophy. Philosophical and theological literature of all faiths, are replete with instances where scientific metaphors are mistaken for proofs, and no distinction is made between scientific and philosophical reasoning.

The importance of the study of scientific paradigms can also be grasped from a slightly different perspective. Even spiritual truth, call it plenary consciousness, God consciousness, knowledge of the Self, etc., will leave behind definite traces in the phenomenal world, if we are careful to observe them. The meaning of trace can be made clear from an example in the realm of scientific inquiry. The relatively recent finding of micro-wave radiation enveloping the universe constitutes a validation of the Big Bang Theory, which is currently the most prevalent theory about the origin of the universe. Since there cannot be a direct proof for the latter, we heavily depend on indirect evidence. After all, it is our experience of partial knowledge that gives rise to the insatiable hunger for complete knowledge. When there is no partial knowledge at all, the question of gaining complete knowledge does not even arise. Consequently, we carefully observe what could be the pointers available in the real world towards the existence of realities in the non-manifest field of existence. Such traces (see 4.11) that are left behind in the phenomenal world, when carefully analyzed, reveal a great deal about the truths of the transcendental realm.

In the Western world, the New Age philosophers who are attracted to Eastern religions lay exclusive emphasis on meditative practices which are also an integral part of the Vedic religion. Unfortunately, this is an incomplete message leading to a lopsided approach. Like every other world religion, Hinduism also emphasizes the need for cleansing the doors of perception through a good moral and ethical life. It is the twin aspects of meditation and moral action that constitute the positive aids for a spiritual aspirant to undertake the journey towards spiritual perfection which, admittedly, is a goal that transcends both Logic and Ethics.

The study of Vedic philosophy and Vedic religion is not just meant to improve one's scholastic knowledge. It is, on the other hand, meant to lead the individual towards spiritual perfection. In the past, we relied exclusively on classical studies for our understanding of the central tenets of the Vedas. However, it is now possible to make these tenets more accessible through a basic understanding of the history of scientific ideas. As expressed in the "Tao of Physics", science does not need religion and religion does not need science, but man needs both of them.

Our main conclusions on spirituality from the Hindu perspective can now be summarized: a) there is no inherent conflict between the belief systems of science and religion since they belong to two different orders of reality; b) the popular interest evinced in mystical experience, particularly arising out of the increasing awareness of mind-body problems, though extremely important, should not, however, be construed as the final goal of the spiritual journey; c) spiritual ignorance, implicit in man's feeling that he is a separate entity, can only be overcome on the basis of the right spiritual knowledge, and, in this respect, the Vedic philosophy has the status of a valid means for acquiring knowledge of the transcendental truths that are quite distinct from those derived by scientific investigation; d) spiritual experiences, very much like worldly experiences, become meaningful only when they are assimilated in terms of right knowledge, and Vedic philosophy offers a systematic methodology for imparting such knowledge based on tradition (*sampradāya*); e) pursuit of spiritual knowledge should not be viewed as other-worldly since it has a definite bearing on enriching worldly life by imparting a proper set of values; and finally, f) there is no claim made for establishing the superiority of the Hindu religion on the basis of the ideas of *exclusivity* and *uniqueness* since these are quite extraneous to the thesis of *sanātana dharma.* On the other hand, there is a strong conviction that all world religions, when properly interpreted, are alternate paths to realize the ultimate truth. The word *religion* is rarely mentioned in the text because Hindu religion is truly a way of life that does not conflict with either science or other religions. This continues to be the mainstream view of the faith, although in practice of late, it has to contend with unfortunate aberrations from its fringes smacking of fanaticism.

Part I

Basic Concepts

Chapter 1

Some Paradigms of Science

The Vedas constitute the principal Hindu scriptures, and their final sections, called the *Upaniṣads*, deal with the important aspect of spiritual knowledge concerning the core truth of human existence. Our aim is to enrich the discussion of the central message of the *Upaniṣads* in light of the philosophical underpinnings of some of the major scientific discoveries. Undoubtedly, the many scientific paradigms from the past to the present have had a decisive influence on the way Western philosophers have thought about the mysteries of nature. Vedic philosophy, however, has not depended upon an understanding of natural philosophy, and as such no background in science was deemed necessary for grasping its significance. Understanding the paradigms of science can, however, sharpen our insights into Vedic philosophy. In addition, that knowledge can also serve the incidental purpose of safeguarding against incorrect statements about scientific truths in the philosophical exposition of the *Upaniṣadic* message. In order to understand the significance of scientific paradigms, we commence with a brief introduction of the philosophical framework that has provided the impetus for the steady progress of the scientific enterprise. Later, we discuss some of the difficult issues that concern science while dealing with the problems that lie at the interface of science and philosophy. The exploration of these issues would call for an understanding of the philosophical import of some of the major milestones in the development of science. Our brief references to the various scientific theories will have the very limited purpose of extracting these principal philosophical ideas.

The frequently–quoted, terse statement made by the seventeenth century French philosopher René Descartes (1596-1650), namely, *I think, therefore I am,* (*cogito ergo sum*), best summarizes the dualistic philosophy based on the separation of *mind* and *matter*, which has guided the scientific view of the universe for a long time. Without the tacit acceptance of this reality, exploration of the scientific laws underlying nature would have been an impossibility. The rationalist viewpoint of Descartes expressed in very few words serves as an excellent point of entry for our later discussion of Vedic philosophy. It allows us to examine the nature of the real 'I' in the Descartes premise and also to comment on its profound implications for nature.

A careful examination of the Cartesian statement about the nature of our existence reveals a hidden paradox. Thinking is a process that is characterized by swift movement. We go from one thought to another with great rapidity and seldom with any control. Each thought that is experienced pertains to a particular state of reality, and the entire process has the transient character of generation and decay. Thoughts come and go at lightning speed. The process of thinking is described as *temporal* because it lacks permanence or stability. In contrast to this, there is also the simultaneous experience of the subjective feeling of 'I', the *individualizing principle*, which remains unchanged throughout the cognitive process. It always remains the same 'I' and as such is characterized by a feeling of constancy; it is thus called the *atemporal* or the spiritual aspect of the twin features of existence. For the present, it will suffice to take note of the simultaneous presence of the temporal and atemporal aspects implied in the Descartes statement that contains the paradox about our own existence which is experienced in the waking state of consciousness.

The temporal aspect of the process of thinking is characterized by time, whereas the atemporal aspect, which is the individualizing principle, is characterized by something that is suggestive of eternality because of its immutability throughout one's life span. The experience of the feeling of 'I' is the same whether one is young or old, whether one is happy or unhappy, and this stubborn notion which lasts a lifetime is quite independent of the physical or emotional state of the human being. The aspect of temporality is called *becoming*, and the atemporal entity is called *being*. The temporal aspect is due to the experience of the passage of time caused by the sequence of past, present, and future events. Because of the subjective experience of the 'flow of time', we can only remember the past, experience the present, and anticipate the future, and this sequence seems irreversible. The future will eventually become the present and therefore is expressed by the terminology of becoming. Being, on the other hand, emphasizes the constancy of the feeling of 'I' which is ever present, unaffected by the flow of time. It is always the present that can be experienced. What we mean by a past experience is only a recollection of the past event at the present moment. Similarly, we anticipate the future at the present moment only. Obviously, there is more to the statement, 'the present is the only reality' than what appears on the surface; there is a deep philosophical meaning to it. The mystery is how the ephemeral experience of becoming, can be rooted in being, which is altogether stationary.

Taking the coexistence of being and becoming at the level of the human mind into cognizance, the next step that the scientist takes is to invest similar properties to our external universe. This is based on the assumption of a direct parallelism between the microcosm and macrocosm: what is valid for the individual on the basis of his actual experience of existence should also be valid to the external world. After all, we notice that there is an absolute interdependency in nature that can be attributed to the concept of constancy and permanence. The sun rises on time, the waning and waxing of the moon has a certainty to it, the fluctuation of seasons is regular, and the planetary motions in our solar system are so predictable that we produce almanacs well ahead of time. The element of constancy of nature, its being, is attributed to the *laws of nature*, whose determination depends on the scientific enterprise.

On the other hand, we also observe that there is a definite impermanence to the objects that we directly observe, and we infer that even those objects that are beyond our direct perception also suffer from the same fate of transient existence. These are inescapable conclusions that we arrive at if our senses are alert and we have recourse to valid means of perception and inference. Within our own species, we see that there is change at every instant from birth to death, following the rhythm of creation, sustenance, and dissolution. One has only to recall that all human cells are replenished once every seven years. Even on the mental plane we experience emotions of joy and grief, pleasure and pain, optimism and pessimism, and many such dualities of feeling which have no permanence to them. As for the objects in nature, there exist distinct entities with their own specificity and uniqueness which we directly observe by their names and forms (*nãma and rũpa*). These are, nevertheless, transitory phenomena. On the basis of scientific reasoning, we know that even the sun will one day in the remote future run out of its fuel. As for the universe, physicists inform us that it too has a finite end in the very remote future counted in billions of years.

Based on our observations and scientific reasoning, we can validate our conjecture of the prevalence of a paradoxical union of being and becoming in the external universe.

In fact, we can attribute a *universal mind* to describe nature's feature of constancy. It is the universal mind along with the transitory feature of becoming that defines nature's existence. The universal mind is represented by the laws of nature, and its twin aspect of becoming is represented by the material universe. There is, however, one important distinction between the paradox of existence at individual and cosmic level. The former is based on the actual experience of the individualizing principle, the 'I' in us. We do not look for further evidence for establishing the truth about our own existence, whereas the corresponding truth pertaining to the universal mind is a matter of speculation and depends solely on observation and understanding of what is observed.

The march of science started with the robust optimism that it was indeed within our grasp to decipher the ultimate meaning of existence for the universe as a whole, quite apart from what is experienced at the human level. It was the riddle posed by the universal mind that became the central focus of the physical sciences, and in that process the search for the core meaning of constancy of the individual mind was set aside for all intents and purposes except for the practical necessity of its role as an *observer* of the external universe. Accordingly, the exploration confined itself to an intellectual inquiry into the ultimate meaning of the laws of nature by determining their common origins. The focus was on ascertaining a theory which could embrace all other prevalent theories of nature. Although there has been remarkable progress in our understanding of nature through physical theories, the core truth of the universal mind has eluded the grasp of the scientists.

We have so far pointed out the manner in which the question of *ultimates* arises as a consequence of the famous Descartes statement *I think, therefore I am.* Before proceeding with our main theme of scientific paradigms, we shall now briefly comment on how a similar question arises within the framework of Vedic philosophy. A detailed comparison between Vedic philosophy and rationalist philosophy is not possible since the origins of their development are entirely different. Vedic philosophy is based on revealed knowledge, and its exposition starts from postulates concerning the *ultimates* of the individual mind and the universal mind. It postulates the existence of an immanent reality, a spiritual element, as the substratum of the individual mind and a transcendental reality as the substratum of the universal mind. The detailed discussion of the immanent and transcendental realities and their interrelationship constitutes the central message of Vedic philosophy. The nature of this philosophical thesis might even suggest that it is completely outside the realm of rational thought altogether. However, rational thought taken to its limits does shed light on Vedic philosophy, and it is in this context that the above discussion based on the Cartesian separation of body and mind becomes useful.

Vedic inquiry also agrees with the notion of constancy attributed to the individual mind and the universal mind in attempting to establish the core truth of existence of man and nature. However, it also points out some basic differences in approach with the result that the meaning of constancy itself gets fundamentally altered. The inquiry proceeds from a meticulous investigation of the individual mind rather than from its counterpart investigation of the universal mind as is the case in scientific investigations. This deliberate choice is based on the supreme assurance that the truth about the existence of the individual mind is at the level of internal experience and hence does not depend on any further validation. The universal mind, on the other hand, is outside the pale of internal experience. It is this initial choice of proceeding from an investigation of the individual mind that sunders the connection between Vedic inquiry and natural philosophy as far as the search for *ultimates* is concerned. This statement is not meant to relegate the importance of scientific inquiry into the understanding of natural phenomena based on the exploration of the universal mind. All that the Vedic philosophy claims is that the search for *ultimates* through the route of the universal mind is fraught with difficulties because it is bereft of any internal experience. Furthermore, each milestone in the long journey has to be validated through objective means of verification which themselves are part and parcel of nature whose ultimate reality is under investigation. We will have occasion to comment on this logical paradox in our later discussion.

Having made the choice of focussing on the individual mind, the Vedic inquiry is quick to assert that it would be wrong to associate the element of constancy with only the waking state of mind as in the case of scientific inquiry. The scope of the inquiry is enlarged to consider the question of constancy, the *being*, within the totality of existence of the individual mind where due consideration is given to all the three states of consciousness, namely, the waking state, the dream state, and the deep sleep state. This analysis of the three mutually exclusive states of mind is not in the domain of psychology, but rather of philosophy. We shall not dwell on this subject further at this stage since it is dealt with more fully in a later chapter.

While exploring the truth about the *ultimates*, we should remind ourselves that we have a wide range of belief systems extending all the way from faith, dogma, indoctrination, and myth on one hand to scientific understanding on the other. The latter is based on formal, logical methods of reasoning and experimentation. In its search for the core reality behind the laws of nature, scientific inquiry uses only the formal methods of reasoning to which it ascribes total objectivity. The empirical scientists believed conclusions do not depend upon the subjective element of the observer. However, we will have occasion to amend this statement later where we have to consider the simultaneous presence of the observer and the observed while dealing with the philosophical significance of the physical theory concerned with the micro world of particle physics.

Credit is given to the ancient Greeks for coming up with the formal methods of logical reasoning as well as originating the study of geometry, which is an important branch of mathematics even to this day. It is interesting that *tarka śāstra*, which is an integral part of the *Nyāya* philosophy of India, is also devoted to the formal methods of reasoning. Since the interchange of philosophical ideas between the West and the East is comparatively recent, it is safe to conclude that these developments were of independent origin. Greek philosophers' formal methods of logical reasoning made use of axioms, deductive logic, inductive logic, analogical reasoning, and concepts of quantification such as numbers, measures and the like. Underlying all this intellectual rigor was the confidence that reality could be grasped by pure reason alone without any resort to experimentation. It is said of Plato (427–347 B.C.), the most influential of all ancient Greek philosophers, that he had a sign posted on the portals of his academy saying that any one who did not understand geometry should not gain entrance. Such was his confidence that geometry could serve as a window into the mysteries of the physical universe.

Plato visualized two entirely different spheres for the concepts of being and becoming. He held that the aspect of being belonged to a transcendental realm, which could be described in geometrical terms. He called them geometrical forms which he termed the Absolute reality. He assigned a lower status to the ephemeral realm which was dependent on and participated in the transcendental realm. This classification and hierarchism is reminiscent of the definitions of *parā vidyā* (higher knowledge) and *aparā vidyā* (lower knowledge) of Hindu thought, which we will explain later. In fact, Plato postulated the existence of two Gods, one having supremacy over the transcendental realm and the other over the temporal realm. In other words, he ascribed one substratum to the element of being and a second to the element of becoming. Furthermore, he considered the world of becoming to be an illusion from the point of view of the transcendental realm. By this he did not mean that the worldly reality is an illusion, rather, that it is illusory when viewed from the vantage point of the transcendental plane. We shall come across a similar but much broader concept, called *māyā* in connection with the Vedic philosophy of nondualism attributed to Śaṁkarācārya. The metaphor Plato gave to illustrate his idea of illusion is very revealing. Suppose, Plato said, we imagine a man in a cave with his back to the light coming from a fire, in which case the objects between the fire and the observer project distorted shadows on the cave wall. The temporal aspects of reality were likened to those shadows, which are quite illusory or merely representations of something antecedent. Although Plato recognized the intrinsic difficulty in reconciling the aspects of being and becoming, he did not offer anything to resolve the paradox. Consequently, the relationship between being and becoming remained a puzzle in his philosophical thinking.

Plato's student Aristotle (384–322 B.C.) put forth an entirely different thesis for this metaphysical problem. His emphasis was more on the aspect of becoming rather than on that of being. Aristotle visualized the universe as a biological organism on par with the human organism. He held that just as the human organism develops according to an unfailing purpose towards a predetermined goal, likewise, the universe is also drawn inexorably towards a predetermined goal. Accordingly, his emphasis was on a teleological purpose for the universe. One can say that Aristotle's thesis has the genesis of the biological concepts that are gradually acquiring prominence in modern day cosmology, which was for a long time dominated exclusively by ideas of physics and a lifeless universe. Specifically, the picture of a transcendental realm painted by Plato did not figure in his scenario. He too, however, paid scant attention to clarifying the paradoxical relationship between becoming and being in the experience of human existence.

Having briefly stated the opposing metaphysical positions of Plato and Aristotle, we shall now present an overview of the development of scientific ideas concerning nature. Our purpose is to focus on the paradigms that science has offered and on the resulting shifts in consciousness over the centuries. This will involve an examination of the philosophical underpinnings of Newtonian mechanics, Einstein's theory of relativity, quantum mechanics, chaos theory, Darwin's theory of evolution, molecular biology, and some other allied subjects.

The constancy of nature, its being, is attributed to the laws of nature. The feature of permanence is all too conspicuous in these laws because they are valid at all times in their own spheres of application. The law of gravity is just as valid in the farthest reaches of the universe as it is on our planet. There is a universality about them, and they are definitely considered to be properties of nature rather than inventions of the human mind. Proceeding from this robust confidence, the scientist is motivated to probe into the mysteries of nature and discover new laws, both through pure reasoning and experimentation, that are subject to the formal rules of scientific methodology. In effect, the scientist conducts a dialogue with nature in order to unveil its secrets, and to further his understanding. At one time in scientific history, there was the implicit faith that this line of inquiry would one day allow us to answer the questions of existence of the universe itself. Some scientists have not yet abandoned the confidence.

There is also another point to be made which is apropos in this context. Unquestionably, since scientific methodology was almost exclusively pursued in the western world, the philosophical underpinnings of the great discoveries of science and the resulting shifts in consciousness had a great impact on the cultures associated with the Judeo-Christian religions. For instance, when the Italian astronomer Galileo (1564-1642) confirmed the Copernican thesis of planetary motion using the new technological innovation of the telescope coupled with the necessary mathematical calculations, he incurred severe penalties from the Vatican. The inhuman punishment he received is incomprehensible to the modern mind. His revolutionary discovery that our planet, revolves round the sun which is fixed in our planetary system was extremely disturbing to the Christian theologians of the day who had, until then, believed in the geocentric arrangement of God's creation. Any theology which depends for its validity on the real facts of nature such as the true picture of planetary motions, the physical origin of the universe, or the origins of life on this planet is bound to suffer the shock waves when its traditional wisdom has to be revised in the light of new scientific discoveries. The principal paradigms of science highlight the sudden changes in consciousness that have occurred, which in turn shed further light on the paradox associated with the conjunction of being and becoming.

Issac Newton, the British physicist of the seventeenth century, is best known for composing the *Principia* in 1687 in which he establishes the three laws of motion. These laws of nature are applicable for predicting the motion of a pendulum, of the planets, and, in principle, of every particle in the universe. The spectacular success of Newton's theory first came in astronomy where it could even account for the existence of Uranus, Neptune, and Pluto, which were unknown before that time. The theory had sweeping implications about the way our external world functioned. In the ultimate analysis, all the events of the past, present, and future were determined by the very first instant in time. The universe behaved like a giant clockwork mechanism independent of human intervention, and there was nothing that was left to chance. The motion of material bodies was guided by a compelling necessity. Newton's equations presupposed a mechanistic interpretation of the universe. Furthermore, the dynamical equations which predict the motion of the material bodies are equally applicable to positive and negative values of time, which means that the theory assumes that time is reversible.

The philosophical implications of Newton's theory can be summarized in the key concepts of *causality, determinism*, and *reductionism*. The law of causality refers to the idea that for every effect there is an antecedent cause. The reasoning behind the notion of the First cause for the origin of the universe stems from the law of causation applicable to a never–ending chain of cause and effects taking us all the way back to the origin of the universe.

By determinism, we mean that once we are given the parameters specifying the behavior of a system at any particular moment, which is called its state, we can determine any future or past state of the system in an unambiguous way. If, for example, we are given the state of motion of a rigid body at any given instant, specified by its position and velocity variables, as well as their initial values, we can determine the values of these variables at any other instant, whether in past or future. The ramifications of Newton's equations for the physical evolution of the universe are interesting. The conditions prevailing at the origin of the universe predetermine the evolution of the universe, leaving nothing to chance. No wonder that cosmological theories that do not ask for the exact information on initial conditions have an intrinsic appeal since, one does not have to contend with the problem of what or who initiated the mechanism. Rather, attention is focussed on the evolution of the universe once it starts evolving. These far–fetched ideas of the universe are still in the realm of speculation.

By reductionism, we mean that the behavior of the system as a whole can be determined once we know the characteristics of the components and their manner of interconnections. For example, the characteristics of an electrical network as a whole can be calculated knowing its various component characteristics together with a knowledge of how the components are interconnected. It boils down to the analysis of a large network through its sub-networks. When the assumption of reductionism is true, as it is in a vast array of problems and disciplines, one is not baffled by the size of the problem since it is possible to start the analysis from its basic building blocks. The underlying analysis of a space shuttle, with its extremely complex system consisting of thousands and thousands of electrical, mechanical, thermal, hydraulic, and chemical components, is based on reductionism. In fact, it constitutes the principal method of analysis for a wide variety of important problems in the field of technology. Reductionism makes it possible to study the mysteries of the universe through a process of segmentation into its component parts, which for all practical purposes are not coupled with each other.

In Newton's mechanistic theory, the assumption is that time is reversible, which is quite contrary both to what we actually experience at the subjective level in terms of the ' flow of time ' and also to what we observe about the creation, sustenance, and dissolution of objects in the external world, including our own human bodies. On the basis of common-sense observations, we arrive at the conclusion that the changes in this world are unidirectional in nature. This unidirectional phenomenon is figuratively expressed as *the arrow of time*, which is oriented from the past through the present to the future. Although the mechanistic interpretation of the universe suggests reversibility because of the way time enters into Newton's equations, our observations confirm that the physical processes are, in fact, irreversible. The contradiction between mechanistic theory and common sense is resolved by the *second law of thermodynamics*, which has profound implications for the evolution of the universe as a whole.

The discipline of thermodynamics had its origin in the study of heat engines. The simple observation that heat can only flow on its own from a hot body to a colder body was critical in recognizing a law which is unidirectional in nature. The consequence of the second law is popularly described by the statement that, in a physical process, with the passage of time, there will be a continual increase from order to disorder until a state of thermodynamic equilibrium is reached. The ideas of order and the various stages of disorder until the process reaches a maximum introduce the ideas of chance and uncertainty. The truth underlying this physical observation is stated more precisely in terms of a quantity, called *entropy*, which is a measure of probabilistic uncertainty or disorder. This word has entered into everyday parlance to describe a variety of disorders that we encounter in various phenomena. Entropy has come to designate confusion of some sort or the other. When the entropy concept is applied to the universe as a whole, we say that the entropy of the universe goes on increasing until it reaches its known maximum value, which represents the state of thermodynamic equilibrium. In fact, it is this important law that enables physicists to conclude that the time will surely come in the far future, again reckoned in billions of years, when the universe will suffer a *heat death*. This phenomenon is the very opposite of the *big bang*, which signifies the event of creation of the universe. The implications of heat death are quite pessimistic. The only consolation to this steady limp towards total annihilation is that it is going to happen in so distant a future that it need not immediately concern us. However, since philosophy seeks to grasp the ultimates of existence, the paradigm suggested by heat death will have a profound influence on our inquiry.

From the above discussion of scientific paradigms, it should be apparent that we would be well advised to give up our insistence that the universal mind can be fathomed through our customary ways of thinking about worldly matters. In fact, most of the difficulties in understanding the abstract ideas stem from our blind devotion to the idea that they are accessible to our common sense. A mental disposition capable of admitting that there could be truths beyond the reach of our common sense is absolutely essential to understand several of the scientific theories, among them Einstein's theory of relativity. With his theory of relativity, Einstein liberated himself from such an idea.

We are familiar with the notions of three-dimensional space and time because of our actual experience with them. Our notions of length, breadth, and height are as intuitive as our experience of the passage of time in terms of the steady sequence of past, present, and future events. But according to Einstein's theory, time is not an absolute and universal quantity because it undergoes deformations with motion. Einstein was primarily concerned with motion close to the speed of light because of his investigation into the nature of gravitational force. Furthermore, space is elastic, and time and space are interrelated. This inextricable relationship between time and space has given rise to the notion of a composite entity called *space-time*, which is a four dimensional quantity (three dimensions of space plus one dimension of time) and which is essential for explaining the curvature of the universe and the force of gravitation. Experimental confirmation of Einstein's theory, which was based on pure reasoning, was provided by Sir Arthur Eddington, the British astrophysicist. The special theory of relativity has altered our view of temporality in a very fundamental way.

One obvious conclusion to be drawn from this paradigm is that the space-time concept is very different from our ordinary notion of 'flow of time' and all that it entails. The ideas of past, present and future have to be revised in the light of the hybrid concept of space-time.

Perhaps the more widely understood result from the theory of relativity is the equivalence of mass and energy as given by the famous equation $E = mc^2$. The knowledge that the physical universe, made up of mass, can equivalently be viewed as energy has given rise to all manner of interpretations by theologians eagerly looking for support of their view that the universe is suffused with divine energy.

It is necessary to recall at this stage that at the time Einstein proposed his theories, the universe was still considered to be static. It was only later that the American astronomer Edwin Hubbell came up with the seminal discovery that we are living in an expanding universe. He also observed that the farther the galaxies are, the faster they are receding from us and that this conclusion is true when viewed from any galaxy in the universe. This discovery led to the formulation of the *big bang theory* which is the most preferred theory to date about the origin of the universe. To put it in rather elementary terms, since the universe is expanding, at some time in the past it must have started from a highly contracted state. This line of reasoning leads to the central idea of the big bang theory: the universe started with an unimaginably big explosion some 17 billion years ago and started expanding uniformly in all directions. As the famous British physicist Stephen

Hawking noted in his widely read book *Brief History of Time* [19], it is only after Hubbell's discovery that the investigation of the origin of the universe became a legitimate scientific discipline. The genesis of nature could now be discussed without any reference to the question of what caused the big bang in the first instance. In other words, the enigma of the First Cause could not have been avoided under the assumption of a static universe as it can be under the possibility of an expanding one.

Hubbell's discovery and the later postulation of the big bang theory were nevertheless, welcomed by many Christian theologians as confirmation of the Biblical truth about the historicity of creation. On the basis of the concept of space-time, it was possible to assign a clear meaning to the aspect of eternality of God. Eternality, in this context, does not mean from everlasting to everlasting. The preferred interpretation, which has a sound scientific basis in relativity theory is that before the event of the big bang, the concept of time is not even defined. In Christian theology, St. Augustine is given credit for holding this view; it is also held by exponents of Vedic philosophy. In order to indicate its eternality, the Absolute is defined as something which is beyond time. From this viewpoint, it is rather impertinent to question what God was doing before creation because it is an ill-defined question and consequently can be brushed aside. The notion of eternality suggested by the concept of space-time is of immense significance in all philosophical discussions irrespective of their origin because the notion of a personal God will necessarily bring about a reconciliation between time and eternity. This, in turn, will have further implications for the concepts of the omnipotence of God, free will, etc. which we will discuss later.

Next, we will comment on what we can infer from the subject of Quantum Mechanics, a discipline that deals with the micro world of particle physics. This theory ranks equal in importance to the theory of relativity in the physics of the twentieth century. Heisenberg's famous *uncertainty principle* puts an end to the idea of strict determinism espoused by Newton and Einstein. His observation, which engendered another paradigm shift is that we cannot simultaneously measure some complementary variables such as position and momenta to a high degree of accuracy. More simply put, such measurements at the micro level are always subject to random fluctuations. For the first time, we have the startling revelation that the microworld is not merely a scaled-down version of the macro world. Consequently, the deterministic models that deal with phenomena such as Newton's laws of motion, Clark Maxwell's electromagnetic theory, Einstein's theory of relativity, etc., are not applicable to the micro world at all.

The concepts of uncertainty and probability have to be invoked to deal with this new situation. It is also important to emphasize that the uncertainty of the situation is intrinsic to the problem and is not the result of our ignorance of a supposedly deterministic order enveloping it. Einstein had great difficulty subscribing to the notion of probability because in his view it was a man-made concept, and he expressed his disapproval in his famous words: "God does not play dice with nature". Despite such expressed doubts, however, it is well-known that the micro-world is characterized by probabilistic uncertainty. This conclusion has profound implications in that it points to the fact that at the atomic level, matter is not at all inert in contrast to our common-sense understanding of nature at the macro level.

This conclusion leads yet again to another important paradigm: our ordinary view of materialism has to be revised to accommodate the fact that at the atomic level, matter can make decisions although the atoms are governed by a statistical law. The pejorative overtones sometimes given to the term 'scientific materialism' seem to have some validity when associated with the concept of inert matter, but have absolutely no meaning when we consider matter at its subtler layers of existence. The distinction between mind and matter gets blurred when we assume that matter is not at all inert at the quantum level as it is at its gross level of existence. (The word 'gross' as used in philosophical discussions means the opposite of the word 'subtle'; specifically, the word 'gross' does not connote the meaning of its modern usage.)

Apart from the non-inertness of matter at the micro level, Quantum Theory has other deep philosophical implications. The demise of determinism also ushers in the end of causality. An effect need not have a cause, which, in turn, means that there is nothing logically wrong if the long chain of cause and effect leading back to the First Cause gets ruptured somewhere. This puts the discussion about the First Cause in a totally different light than in deterministic theories. Furthermore, there is no clear separation between the observer and the observed. Accordingly, consciousness enters into the picture of observations which means that we cannot visualize a universe divorced from man. This presents a *holistic* picture of the universe, which states that everything is connected to everything else in this universe in a gigantic network of relations. Quantum Mechanics admits the interpretation that the whole universe is in flux, a concept that is central to Eastern philosophies.

Most of the cosmological theories that have been advanced so far are purely physical in nature and do not take into account the possible effect of life on those theories. This, according to some renowned scientists, is a grave omission since the effect of life may indeed result in an altered view of our cosmic future. The appearance of life on our planet is a relatively late phenomenon in the history of the universe. Estimates of its exact time vary, but according to one account, it started around 5 billions years ago if the big bang occurred 17 billion years ago. In one sense, while the physical theories have looked into the past of the lifeless universe, the phenomenon of life has greatly influenced what will happen to the universe in the future because of its undeniable presence, at least on this planet. There is a great deal of interest in investigating whether life exists in other parts of our galaxy, which is also an interest encouraged by the possibilities of space travel. There is also the distinct possibility of planting life on a lifeless planet thanks to modern day technology. These are compelling questions, but we shall not dwell on them because they are not directly related to our present metaphysical investigation.

Undoubtedly, Darwin's theory of evolution has had a profound impact on Christian theology. Even to this day, there is an ongoing debate between the creationists, who have faith in biblical exegesis, and the evolutionists, who endorse Darwin's Origin of Species. The contentious point about the biological theory, which to our immense relief is within the intellectual reach of a substantial number of people because of its reliance on concrete concepts, is that evolution is attributed to the principle of random mutations and natural selection, to *chance and necessity.* Taking a simple example, an ensemble of colored polar bears in a snow-clad landscape would be immediate targets for its predators because they would be all too conspicuous. If by pure chance, one white polar bear appears on the landscape, its progeny of white polar bears will ultimately survive in the snowy habitat and will eventually lead to the total extinction of the colored ones. Darwin's work is based on actual observations and voluminous data collected over years of painstaking, scholarly work. His is not a theory directed towards unification of phenomena, as in theory of relativity or quantum mechanics. Rather, it is an effort to bring some meaningful consistency to the phenomena of life on this planet.

From Copernicus we learnt that our planet revolves around the sun, a fact which was later experimentally confirmed by Galileo. This discovery came as a blow to the egocentricity and faith of man living on this planet. In Darwin we encounter the idea that the anthropoid ape was an ancestor in our long evolutionary history, the knowledge of which inflicted yet another blow to man's pride. Those who believe that man has been placed on the face of the earth with a divine purpose without reference to any other species reject the theory of evolution, but as it is based on the uncompromising honesty of science, we seem to have no choice but to accept the fact grudgingly. The Hindu belief of a fixed number of species since creation with man at its pinnacle also conflicts with the scientific view of evolution.

The physicist Schroedinger, who was working in the relative serenity of Dublin, Ireland, during and immediately after the Second World War, gave a series of lectures, from which his famous monograph, *What is Life* stemmed [41]. He was a renowned quantum mechanician, who knew of the vital importance of the emerging discipline of biology. He was careful enough not to raise the question of the origin of life in his lectures since he deemed it too premature to bring up that question in any meaningful way. Instead, he concentrated his attention on the question of life once it appeared on this planet.

Scientists in life sciences have given a great deal of credit to Schroedinger for having provided the timely inspiration for the development of the modern discipline of molecular biology; references to RNA, DNA, double helix, genetic engineering, and the like have now become commonplace. This amazing discipline has, to a large extent, cracked the secret code of life, and the frenetic effort has resulted in the complete map of the human genome. We are already witnessing the beneficial impact of the discipline in medicine and many other areas, even during this relatively early stage of scientific research and development. One can clearly state that, in philosophical terms, molecular biology is based on reductionism: a total picture that emerges from the basic building blocks at the molecular level. The holistic view of life, on the other hand, denies the cogency of decomposition. The discipline of molecular biology has reached such an advanced stage that the technology may soon be available for cloning a specific creature. Claims of successful human cloning, as yet unverified, have already started pouring in. Because it would take us away from our main theme, we shall not comment further on the ethical questions associated with this kind of work, which, at the cutting edge of biotechnology, is fraught with morbid fear and an acute concern for the future of the human race.

Apart from the biological basis for life, which is an important question in itself, the further question that is of interest to us in connection with our broader understanding of the metaphysical problem is the manner in which consciousness has come into existence. It is consciousness and self-awareness that characterize the spiritual element in man. Scientific opinion on the subject generally acknowledges that it is beyond the reach of molecular biology alone to come to grips with this intractable question. It is speculated that the emerging discipline of 'Chaos' (an unfortunate name for a discipline which has so much promise) may contain the clue because it presents a coherent theory of self-organization, which is germane to the study of consciousness.

Chaos is a theory that deals with non-linear phenomena where the relationship between cause and effect does not have a strict proportionality as in the linear case; even incremental change in a causative factor can bring about a disproportionate amount of change in the resulting variable. Examples of this phenomenon are ubiquitous in nature.

The so-called 'butterfly effect' has now become the current metaphor to illustrate the principle: flapping wings of a butterfly in a distant city can produce drastic changes in the weather patterns elsewhere. In other words, the change is extremely sensitive to initial conditions. Some examples other than from meteorology, in which the discipline first made its mark, are seen in a leaky faucet where water drips in a nonuniform way, the normal rhythm of the heart that can suddenly go haywire into an arrhythmia, and turbulence in fluids. The most interesting feature of this phenomenon is that once a disturbance takes place, it will eventually reach a new state of equilibrium which is anything but chaos as the word is conventionally understood. It results instead in beautiful structures undreamt of before the disturbance took place.

The Nobel Laureate Prigogine has called the resulting structures *dissipative structures*. This ability to acquire a new state of equilibrium after experiencing a chaotic disturbance, is in short, the essence of *self-organization*. It is also described as *organized complexity* since it is not caused by uncertainty of any sort that is characteristic of randomness in phenomena that are truly probabilistic in nature, but by clearly-known deterministic reasons, albeit of a non-linear nature. Since the phenomena are non-linear, the net effect of the system as a whole cannot be predicted by the known characteristics of its component parts. This is why we say that the study of consciousness is beyond the reach of molecular biology, which is characterized by reductionism, whereby a system's behavior can be predicted on the basis of the behavior of its component molecules. The emphasis therefore is on *holism* rather than a segmentation of the problem. The non-linear system has to be studied in its entirety because of the impossibility of basing its analysis on its subsystems.

This concept of wholeness with regard to consciousness conforms to our quantum mechanical understanding that the universe as a whole is in a state of continuous flux. The scientific surmise is that this new theory of chaos enables us to hypothesize that consciousness is not a result of the properties of the molecules but of the manner in which they are arranged, namely, on its organized complexity. What is of real significance is the architecture of the molecular arrangements rather than the properties of molecules. For the present, explaining consciousness by means of chaos theory remains a matter of speculation although there is some degree of plausibility to it. The full potential of the theory of chaos, which is based on holistic principles, is yet untapped.

One extremely interesting example of self-regulation is contained in the *Gaia* hypothesis proposed by the maverick British scientist James Lovelock. Gaia is the name of the Greek goddess of Earth, and the hypothesis deals with our own planet. It stresses that life on Earth has definitely modified the environment, and, in turn, the host of physical factors at play have had a definite impact on life on Earth. These two sets of forces are mutually coupled in such a way that the whole planet is acting like a self-regulating system. For now, the Gaia hypothesis is only providing a heightened awareness of the complex self-regulating system working to sustain life on this planet. Consequently, it also suggests an intellectual motive for taking our responsibilities for preserving our ecology more seriously.

Physicist Paul Davies who is extremely optimistic about the rich possibilities of the new theory of chaos makes the following encouraging comments against the background of thermodynamics, which, as we have mentioned above, predicts that the universe is doomed to meet with heat death because of the steady rise in entropy. In 'God and New Physics' [14], he makes the following claims:

> There exists alongside the entropy arrow another arrow of time, equally fundamental and no less subtle in nature. Its origin lies shrouded in mystery, but its presence is undeniable. I refer to the fact that the universe is *progressing*–through the steady growth of structure, organization and complexity–to ever more developed and elaborate states of matter and energy. This unidirectional advance we might call the optimistic arrow, as opposed to the pessimistic arrow of the second law.
>
> There has been a tendency for scientists to simply deny the existence of the optimistic arrow. One wonders why. Perhaps it is because our understanding of complexity is still rudimentary, whereas the second law is firmly established. Partly also, perhaps it is because it smacks of anthropocentric sentimentality and has been espoused by many religious thinkers. Yet the progressive nature of the universe is an objective fact, and it somehow has to be reconciled with the second law, which is almost certainly inescapable. It is only in recent years that advances in the study of complexity, self-organization and cooperative phenomena have revealed the two arrows can indeed coexist.

We shall now turn our attention to the role of mathematics in the study of nature. In the formulation of scientific theories, mathematics plays an important role because of its uncanny ability and versatility in describing natural phenomena, and hence it is aptly described as the queen of the sciences. In fact, some of the mathematical theories from which we have extracted the significant scientific paradigms are considered to be discoveries in themselves rather than mere inventions. A simple illustration of such a discovery in mathematics is the concept of Euclidean space, which is the three dimensional space of our ordinary day-to-day experience. Although it is a geometrical result that belongs to the realm of mathematics, it has all the claim of a discovery because of its intrinsic truth about an important aspect of nature in general. It conforms to our commonsense experience of space. Similarly, there are several results in mathematics in which where the proposed model or theory serves not only the immediate purpose for which it was invoked but also a purpose not originally envisaged.

The question of whether mathematical reality is an invention of the human mind or a true discovery is, however, a matter of controversy amongst mathematicians, but no one can remain unimpressed by some of the superb branches of mathematics which have unveiled the mysteries of nature. The stories of some of the great mathematicians which tell of how they hit upon their great contributions as a result of sudden intuition rather than as a product of scientific reasoning add credibility to the idea of discovery. Rāmānujam, the Indian mathematical genius, who made prodigious contributions to number theory, has claimed that his inspiration came from his devotion to his favorite goddess. It is beyond imagination how he could have thought of so many theorems, which have engaged the attention of mathematicians even to this day, without the aid of some intelligence beyond the reach of formal methods of reasoning. There are several such instances in the history of mathematics which make us wonder about the potential of the human mind.

Roger Penrose, an eminent mathematical physicist from Oxford University has written a classic called *The Emperor's New Mind* [31]. His thesis raises serious questions about the prospects for 'strong artificial intelligence', which is a topic of computer science. The proponents of this discipline claim that the day is not far off when computers will have the ability to do whatever a human mind is capable of doing. Penrose's thesis is quite extensive and includes a majestic sweep of ideas concerning the development of physics and mathematics:

> How 'real' are the objects of the mathematician's world? From one point of view it seems that there can be nothing real about them at all. Mathematical objects are just concepts; they are the mental idealizations that mathematicians make, often stimulated by the appearance and seeming order of aspects of the world about us, but mental idealiza-

> tions nevertheless. Can they be other than mere arbitrary constructions of the human mind? At the same time there often does appear to be some profound reality about these mathematical concepts, going quite beyond the mental deliberations of any particular mathematician. It is as though human thought is, instead, being guided towards some external truth–a truth which has a reality of its own, and which is revealed only partially to any one of us.

After discussing some branches of mathematics in support of his position, he further states: These are the cases where much more comes out of the structure than is put into it in the first place. One may take the view that in such cases the mathematicians have stumbled upon 'works of God'.

We will end our discussion of paradigms from the scientific literature with a mathematical result that has far-reaching implications for metaphysics. It is called *Gödel's incompleteness theorem*, which is interpreted by metaphysicians to mean that even the limits of rational thought cannot uncover the truth of the transcendental realm. For the present, we can define the transcendental realm as where time is not defined. As the physicist Max Planck has put: Science cannot solve the ultimate mystery of Nature. And it is because in the last analysis, we ourselves are part of the mystery we are trying to solve.

Gödel's theorem, of course, is a rare gem in pure mathematics. It is proved rigorously with great ingenuity and resolves one of the basic paradoxes faced in logic and mathematics, that is, the problem of self-reference. An example from logic is the so-called liar's paradox. Consider the proposition, 'this statement is false'. The paradox inherent in it arises because the statement refers to itself, which is an impossible situation akin to that of a man standing on his own shoulders. If the statement is false then it is true, and vice versa, so it is not amenable to the formal methods of logic. We find several paradoxes like these in the mathematical literature.

Mathematical interest in the liar's paradox has a long history. To recount the story briefly, David Hilbert, another mathematician of great fame, listed several unanswered questions in mathematics towards the end of his career. One of them had to do with the possibility of proving all mathematical truths in a formal way from given sets of axioms so that the procedure could be automated. It was this problem that engaged the attention of Gödel, a German mathematician who, after emigrating from Europe, was a contemporary of Einstein in the Institute of Advanced Studies in Princeton. He proved his theorem in the year 1930, when he was very young. The essence of his findings was that there will always remain some mathematical truths which will escape the net of a given set of consistent axioms and formal methods of deduction from them. If one were to enlarge the set of axioms to comprehend the results which have escaped the net in the first try, even at the second stage there will be a further set of mathematical results that will escape the purview of the new set of axioms. This will persist *ad infinitum.* Gödel's theorem is well known not only for its final result but also for the ingenious proof that he devised. Roger Penrose humorously remarked that an interesting consequence of Gödel's theorem which is of tremendous assurance to the profession of mathematics is that the research has no end in sight. In passing, it is interesting to note that Gödel was greatly inclined towards mysticism and spent his last few years in Princeton in prim aloofness.

The problem of self-reference is encountered in several disciplines other than logic and mathematics. Hofstadter's book *Bach, Escher and Gödel* [22] provides an excellent treatment of the concepts of self-reference as they appear in music, painting, and mathematics. The philosophical truth that arises from the phenomenon of self-reliance as it is manifested in these and other disciplines is that even the limits of rational thought and finite methods of analysis are totally inadequate to resolve the mystery underlying the metaphysical dilemma. This conclusion may appear as an anticlimax to our earlier discussion where the focus was on scientific paradigms which had a definite impact on philosophical thinking. However, pessimism is unwarranted since it is precisely the scientific thinking that has sharpened our insights and greatly enhanced our knowledge of the interface areas of science and metaphysics. Even the clear understanding of how far one can proceed with scientific thinking is a great step forward.

As noted previously, Judeo-Christian theology has always been intertwined with natural philosophy. We shall cite a few instances of how the scientific paradigms brought about shifts in the consciousness of Biblical scholars. We have already mentioned the effect felt by the works of Copernicus and Galileo with regard to planetary motions. In the face of compelling scientific discoveries, the church had to eventually relent on some of its dogma, for example, that the sun, along with the other planets, orbit around our earth. On the other hand, science has also supposedly confirmed some of the long-held Biblical truths, for example, that creation began at a definite time in the past. Furthermore, the creator and the universe are considered to be quite distinct since God is believed to have created the universe *ex nihilio.*

The claim for support of this Biblical truth is traced to the theory of relativity which spells out interrelationships between space, time and matter. According to the physical theory, time and space have meaning only after the event of the big bang. Before the event of creation, these entities are not even defined. God is eternal in the sense that he is beyond the description of time, which is a concept from mathematical physics that what lies beyond the event of the big bang is totally inaccessible to historical ideas of time. We have no intention here to get into a detailed discussion of the God of the popular religion and the manner in which the interrelationship of time and eternity have been reconciled in its theology. The following quotation from Joseph Campbell from his *The Masks of God* [5] regarding Christian thought will assist us in pointing out the differences from Hindu thought which we will be discussing later.

> ... In the Bible, God and man, from the beginning, are distinct. Man is made in the image of the God, indeed, and the breath of God has been breathed into his nostrils; yet his being, his self, is not that of God, nor is it one with the universe. The fashioning of the world, of the animals, and of Adam (who then became Adam and Eve) was accomplished not within the sphere of divinity but outside of it. There is, consequently, an *intrinsic* , not merely *formal*, separation. And the goal of knowledge cannot be to *see* God here and now in all things; for God is not in things. God is transcendent. God is beheld only by the dead. The goal of knowledge has to be, rather, to know the *relationship* of God to His creation, or, more specifically, to man, and through such knowledge, by God's grace, to link one's own will back to that of the Creator.

We shall now turn our attention to some of the difficult issues that lie in the grey area between science and Christian theology. Freeman Dyson, the Princeton physicist, has composed *Infinite In All Directions* [18] after delivering a series of Gifford lectures at Aberdeen, Scotland, in 1985. He concentrates on the origin of life, the human experience of free will, the prohibition of teleological explanations in science, the argument from design as an explanatory principle, and the question of ultimate aims. He has dealt with these topics from the point of view of both science and Christianity. The discussion of these interesting problems will also have relevance for our own theme of Vedic philosophy which we are taking up from the next chapter onwards.

As for the origin of life, Biology insists that life originated through a process of chance, whereas theology informs us that life arose as a result of an act of God. The famous French biologist Jacques Monod, who is the author of the book *Chance and Necessity* [28], would have us believe categorically that God's plan has no place in the scientific scheme of things because life made an accidental appearance in this universe. This is also the stand taken by most modern biologists. But the confidence with which they make their assertions can be shaken by the question whether consciousness and self-awareness also appeared as a result of pure chance. The following argument can also be made: chance is purely a human concept, and at best it is a cover for our ignorance about the true state of affairs. Chance in this context comes under the classification of *deterministic uncertainty* as opposed to probabilistic uncertainty that we encountered in connection with quantum mechanics.

According to deterministic uncertainty, the underlying layer of truth can always be unveiled if we can possibly remove the overbearing layer of uncertainty caused by ignorance. In other words, we do not concede that the phenomenon is essentially random when there is no possibility whatever for the uncertainty to be resolved through deterministic methods of analysis. Even explanations in particle physics based on probabilistic uncertainty were opposed by no less a scientist than Einstein when he made his famous comment that "God does not play dice with nature". We know now that we would be on weak ground making this argument because there could be intrinsic random phenomena governed by chance and probability which are subject to statistical laws. Reconciliation of the two opposing views of deterministic and probabilistic uncertainty are next to impossible in regard to the manner in which life appeared on this planet.

Free will is another mystery which concerns both science and religion. Free will, or predestination, is inextricably linked to the human experience of having a separate individuality, one's ego sense. It is a direct consequence of the individualizing principle of ' I, ' to which we made reference earlier. Expositors of Hindu religion refer to this concept with a touch of drama by saying that a human being possesses both the elements of a beast and an angel, and it is up to the individual to exercise his free will in choosing either as his model of action. In Christianity, the religious dilemma is posed in the setting of the one Creator and many separate individuals. The conflict arises because we cannot, on the one hand, vest the qualities of omnipotence and omniscience to the Creator and yet, on the other hand, allow individuals to exercise their free will. If the former is true, then God has complete foreknowledge and control over the future course of individuals, thus stripping them of their discretion. Conversely, if individuals retain their powers of decision, then, the Creator would be denied his absolute qualities of omnipotence and omniscience. This riddle has demanded the attention of theologians and philosophers who proceed from their monotheistic conception of God.

According to Dyson, both narrow-minded scientists and theologians have denied the existence of free will. Scientists such as Jacques Monad deny on the basis of their hypothesis of chance and necessity in matters concerning life on this planet. The denial of the existence of free will by orthodox Christian theologians is based on their emphasis on preserving God's powers of omnipotence and omniscience. In this connection, it is interesting to note that free will is also denied in Hindu thought, but for an entirely different reason. The denial is based on mystical experience, that is, the reality experienced in a higher state of consciousness where there is no experience of time. When one is entrenched on that plane of consciousness, there will be no experience of free will. Time, the chief characterizing factor in the material world, is absent in the transcendental realm. Therefore, the problem of free will is reduced to the incomprehensible relation that exists between time and eternity. Consequently, the conflict inherent in free will becomes irrelevant when viewed from the transcendental point of reference. This is the view expressed by the Hindu sages who are generally credited with the experience of mysticism. In *The Teachings of Bhagavān Śrī Ramaṇa Maharishi*, edited by Arthur Osborne [27], there is a passage on the problem of free will in which one of the greatest sages of India has endorsed this view. Interestingly, the Hindu view would coincide with the view of orthodox Christian theology if the concept of God could be equated with the plenary consciousness of a mystic. We do not, however, want to get embroiled in the purely theological question any further, particularly because it requires a more detailed understanding of the two religions.

The denial of free will, however, will not give us a satisfactory answer to the quandary of teleology versus causality. Dyson proposes a very interesting way out of this impasse: 'Free will is the coupling of a human mind to otherwise random processes inside the brain. God's will is the coupling of a universal mind to otherwise random processes in the world at large.'

Dyson attributes his explanation of free will to a branch of Christian theology called Socinian theology. While it seems an elegant explanation, it raises some further confusion by establishing a coupling between the individual mind and the universal mind, both of which are deterministic in nature and are governed by purely random processes marked by probabilistic uncertainty. The solution he offers seems to be as difficult as the problem that was posed.

Next we deal with the problem of teleological assumptions that do not come under the strict purview of science. We have already encountered such an explanation with regard to Aristotle's view of the universe. He subscribed to the idea that the universe behaves like a biological organism drawn towards a predetermined goal with a final purpose in mind. Such explanations are, by and large, beyond the boundaries of scientific thinking since they attach significance only to causes which are more immediate and local in nature. Perhaps the most important explanation based on teleological reasoning is the anthropic principle.

The anthropic principle has a coveted place in the study of cosmology because of its bearing on the importance of man as the observer. Before we explain what this principle is, it is useful to recall some salient features of our historical universe in order to appreciate the sheer magnitude of our universe and the insignificance of man in comparison to it. Arthur Eddington, the British physicist, has given us a rough estimate of the immense expanse of our universe. One hundred billion stars make a galaxy, and one hundred billion galaxies make a universe. Our own particular galaxy is called the Milky Way, which, according to Eddington, contains about one hundred billion stars. The expanse of our galaxy can be judged by the fact that it takes almost 100,000 years for light traveling with a velocity of 300,000 kilometers per second to cross it. There are about a billion places in the Milky Way which have the possibility of supporting life. Considering all the other galaxies, there are a billion times a billion places which can support life! The significance of the anthropic principle should be understood in this context because it comments on the uniqueness of our own planet and galaxy and the importance of man despite his infinitesimal role as a speck of cosmic dust.

The anthropic principle, which has a great significance in cosmology, amplifies the metaphysical paradox concerning the conjunction of the observer and the observed. The principle arises from the commonsensical idea that it is only the existence of man that has provided the scope for raising fundamental questions about the mysteries of the universe. The main assertion of the principle stems from the physical fact that if the dimensionless constants of the universe (velocity of light, ratio of proton and electron masses, etc.) had been different by even very minute amounts, man could never have come into being. Thus, the universe could easily never have been known. The values of the dimensionless constants making man's existence possible in this universe are extremely sensitive. From this point of view, not only has man adapted to the universe but also, conversely, the universe has adapted to man. To quote Barrow and Tippler [1]:

> The possibility of our own existence seems to hinge precariously upon these coincidences... Our picture of the universe and its laws are influenced by an unavoidable selection effect–that of our own existence... The anthropic principle shows that the observed nature of the universe is restricted by the fact that we are observing this structure; by the fact

that, so to speak, the universe is observing itself. The anthropic principle therefore is a direct consequence of our own existence and enters either explicitly or implicitly into the various theories about our cosmos.

George Wald, a Nobel Laureate in medicine, recalls a witticism he heard that summarizes the essence of the argument of the anthropic principle [42]: "Why is the world five billion years old?". "Because it took that long to find that out!". Thus, the anthropic principle is informed by teleology.

If we do not accord the status of a scientific principle to the anthropic principle on the grounds that it is guided by notions of final purpose, then the onus is on us to seek a satisfactory explanation for what we mean by purpose as experienced at the human level, and also to explain our implicit faith in a purpose for the universe at large. The answer to these perplexing questions seems to lie in correctly recognizing the scope of scientific inquiry. The choice of the laws of nature and the choice of initial conditions of the universe, which are the dimensionless constants that come under the domain of the anthropic principle, are clearly beyond the realm of science. In fact, they belong to meta-science. The jurisdiction of science is only to explain the phenomena of nature without reference to final purposes. Nevertheless, science does not deny the importance of teleological reasoning, rather it asserts that the latter constitutes a complementary style of reasoning that can be called philosophical reasoning.

The argument from design is another philosophical problem which lies at the interface of science and religion. In all arguments from design, one recognizes the twin ideas of a predetermined architecture and a transcendental purpose. The earliest design argument in the history of Western philosophy was presented by Paley, who argued that just as a watch with its intricate mechanism implies the existence of a watch maker, the design of the universe implies the existence of God. Those such as Bertrand Russell, who do not believe in the argument from design, find it eminently suitable for parody. To quote Russell [40], "You all know Voltaire's remark, that obviously the nose was designed to be such as to fit spectacles."

The argument from design, however, engaged the attention of both the creationists and evolutionists of the nineteenth century. The Darwinian theory of evolution brought a halt to the controversy since it did not have to resort to teleological reasoning. The concept of random mutations and Darwinian selection were entirely sufficient to explain the evolutionary process. The argument from design becomes contentious only when we attempt to accommodate it within the framework of science. On the contrary, when it is accorded its rightful place as a philosophical principle, it enjoys the same status as the anthropic principle, distinct from, but complementary to, the scientific principle.

Dyson suggests three levels of mind in the following ascending orders: the level of matter at its subtlest as in quantum physics, the level of the human mind, and the level of the universal mind. We know that matter at the quantum level is not inert but is an active agent making choices subject only to laws of probability. As for the human mind, Dyson [18] states that "our brains appear to be devices for amplification of the mental component of the quantum choices made by the molecules inside our heads". We have already encountered the phenomenon of random processes in the brain when we considered the problem of free will. The argument from design appears at the third and final level of the operation of mind. The laws of nature suggest that the universe as a whole is hospitable to the growth of the mind which, in turn, suggests the existence of a universal mind. Earlier we projected the atemporal aspect, the element of constancy of the universal mind, onto the laws of nature. We drew a parallel between the individual mind and the universal mind on the basis of their twin features of being and becoming. Extending this parallel one step further, we can postulate that just as the individual mind is constantly seeking to determine its own true nature, the universal mind is also vested with the capability for its own self-discovery. In religious terms, we can call the substratum of the universal mind 'God', in which humans become tiny outlets for serving His purpose on this planet.

The concept of final aim to the universe constitutes the last grey area that we are considering here. To explain this concept, Dyson employs the *the principle of maximum diversity*, which can be viewed as an extension of the anthropic principle and the argument from design. To quote Dyson [18]:

> The principle of maximum diversity operates both at the physical and the mental level. It says that the laws of nature and the initial conditions are such as to make the universe as interesting as possible. As a result, life is possible, but not easy. Always, when things are dull, something new turns up to challenge us and to stop us from settling into a rut. Examples of things which make life difficult are all around us: comet impacts, ice ages, weapons, plagues, nuclear fission, computers, sex, sin and death. Not all challenges can be overcome, and so we have tragedy. Maximum diversity often leads to maximum stress. In the end we survive, but only by the skin of our teeth.

In summary, this chapter has defined the central problem of metaphysics, which is the exploration of the ultimate basis for the individual mind as well as that of the universal mind. The existence of the latter was postulated on the basis of the parallelism between the two. The laws of nature with their element of constancy provide the clue to establishing their correspondence with the human experience of the element of constancy implicit in the 'I' notion. The direct concern of science all along has been to explore the mystery of nature and thereby discover the laws that explain phenomena in the universe. This is the so-called *theory of everything.* In its long history of impressive achievements towards realizing its goal, science has hitherto paid scant attention to the mystery that is experienced at the human level, despite the ongoing research on the physics of the brain. The search at the human level is the domain of philosophy and religion. Although both science and religion are interested in exploring the ultimate meaning of existence, science has chosen to explore the mysteries surrounding the universal mind; whereas religion is engaged in the same enterprise at the level of the individual mind. This is the clear point of bifurcation between the two enterprises.

This chapter has presented an overview of the scientific paradigms arising from the great milestones in physics and biology in order to portray their impact on philosophical thinking. Newton's equations of motion introduced the concepts of causality, determinism, and reductionism. Time was entirely reversible in that discipline, which was contrary to our subjective feeling of the chronological flow of time. In thermodynamics, on the other hand, we found that time was unidirectional and the physical processes irreversible, a fact which was embodied in the second law. The dire prediction of this law due to the steady rise in entropy was that the universe was inexorably grinding itself to a heat death in the remote future measured in astronomical units of time. With Einstein's theory of relativity, we had to adapt our concepts of time and space to the notion of space-time. Einstein also concluded that space, time, and matter were all related and that there was an equivalence between energy and matter. From quantum mechanics, we learnt that the micro world behaved in a probabilistic way, which led to the startling conclusion that matter was not inert at that level.

Heisenberg's uncertainty principle put an end to ideas of strict determinism at the particle level thus challenging the strict duality between mind and matter asserted in the Descartes' statement, ' I think, therefore I am'. Human consciousness became an integral part in the process of observing the universe. This integration led to ideas of holism as opposed to reductionism, by which the universe was conceived as a gigantic network of relations rather than something that could be segmented into smaller parts, as studies based on determinism would do.

From Darwin's theory, we noted the role of random mutations and natural selection in the evolution of life. We also commented on how molecular biology has satisfactorily explained the chemical mechanism for the formation of life. From the new discipline of chaos theory, we learnt how new structures were formed once the equilibrium of the preceding structures were disturbed in a non-linear way. This theory has the remote potential of explaining consciousness, an attainment which is presently outside the scope of molecular biology. At present, the latter discipline can only successfully explain the phenomenon of life based on deterministic models. Turning to mathematical theories, we mainly concentrated on Gödel's incompleteness theorem, which problematizes the concept of self-reference and implies that the transcendental realm cannot be accessed even when rationality is taken to its very limits. We have referred throughout to the results of cosmological studies because of their bearing on the search for the core truth about existence. We have also commented on what life sciences have to offer in the explanation of this truth.

The background material of this chapter is essential for the understanding of Western philosophy. As for the understanding of Vedic philosophy, they provide deep insights quite distinct from the insights provided by the classical treatment of the subject. Hindu philosophy starts by investigating the mystery surrounding the individual mind rather than the universal mind, and proceeds on the assumption that both domains of inquiry will lead us to the same conclusion. The choice of this approach is dependent on the fact that while the operation of the individual mind can actually be experienced, the universal mind, on the other hand, lies only within the realm of intellectual knowledge and is bereft of any actual experience with it. Furthermore, only the individual mind does not require any proof for its existence. Evidence for the existence of the universal mind is acquired only through formal methods of logic. However, we shall see that the investigation into the truth behind existence at the level of the individual mind would be incomplete if we cannot bring the infinite character of the universal mind into the picture and make use of the parallelism that we have defined between the two levels of mind. Vedic philosophy, which we will develop in the succeeding chapters, addresses itself to this question in a direct way.

Chapter 2

The Vedas and The Vedic Philosophy

2.1 Vedas

The Vedas are the holy scriptures of the Hindu religion and are considered to be revealed knowledge. Their scope is quite extensive ranging from secular to spiritual matters. The central theme, however, is to declare the existence of the one single eternal transcendental reality that is the 'divine Ground' of both humans and the universe, and to beckon human beings to aspire to its realization in order to lead a fully integrated life. There are several schools of philosophy within the Vedic fold that are devoted to the articulation of this central message, and the one we call Vedic philosophy presents a global overview before we take up a more detailed study of some other allied schools. Our outline is to first present a brief background of the Vedic scriptures and then follow with a comprehensive summary of the Vedic philosophy based on them.

For our exposition of the part on the Vedas, we have relied on some relevant descriptions contained in the book *The Vedas* [38] by the late Śaṁkarācārya of Kānchi Kāmakoti Pītham who is considered to be one of the greatest sages of modern India. We have followed this course in order to preserve the authenticity of the orthodox interpretation notwithstanding our own interspersed comments.

Vedas are referred to in the Sanskrit language as *śrutis.* One distinguishing feature of the Vedas from the scriptures of other religions is that they do not have an author, a fact that is puzzling to the modern mind. They are called *apouruṣeyam* which means not authored by any human being (*puruṣa*). This is in stark contrast to the holy books of other great religions where there is historical evidence of their sources. For Buddhists it is Dhammapada; for Christians, the Bible; for Muslims, the Koran; for Parsees, the Zend Avesta; for Sikhs, the Grantha Sahib; etc. The prophets of these religions are also well-known and revered by followers of their respective faiths. Hinduism is different in this respect since it does not have a single prophet and does not have an organized church. The primary focus of the Vedas is to bring the declaration of the ultimate truth concerning Man and the universe to the attention of those who are interested in treading the spiritual path and also to prescribe the ways of life that are necessary for making progress in the direction of its realization. It is a message that is addressed to the individual spiritual aspirant and takes into account the fact that not everyone will be interested in heeding it. In practice, it encourages respect for all religious faiths because, in the final analysis, they are also interested in fathoming the ultimate truths concerning man and the universe. It is precisely this viewpoint that paves the way for the universality of the message, in the sense that a person belonging to any religion can examine the value of the Vedas to him without prejudice to his own faith. Hindus consider the Vedas as *anādi*, that is, without a beginning. Since anything that is defined in time has a beginning, *anādi* has to be understood as something beyond time. The claim of absence of historicity to the scriptures is difficult for the modern mind to accept, but we shall briefly dwell on the background to this concept. This takes us to the Hindu view of cosmology.

Hindus subscribe to the concept of a pulsating universe with a continuous rhythm of creation, sustenance and destruction. It is not only cyclic, but each cycle has sub-cycles within it. The cyclical model is reminiscent of the model suggested by the Russian meteorologist Alexander Friedman as a solution, within the framework of Einstein's relativity theory, to the origin of the universe. The notion of sub-cycles is specifically Hindu. The duration of the universe can be inferred from the the scriptures on the basis of Hindu astrology (*Jyotiṣa*) which is a sub-discipline within the Vedas. The astrological prediction of the duration of the universe is based on the length of the sub-cycles, the four *yugas*, which also determine the length of the pulsating cycle. Our present yuga, *Kali-yuga*, is supposed to be 432,000 years in duration, *Dvāpara-yuga* twice that duration, *Tretā-yuga* thrice that duration, and, finally, *Kṛta-yuga* four times *Kali-yuga*'s duration which is equal to 1728,000 years. The four yugas put together, is called a *mahāyuga*, the duration of one cycle. One thousand such mahayugas make the period of reign of the fourteen *Manus* who have dominion over this universe. The same period constitutes a day for *Brahmā*.

The night of *Brahmā* is also equal in length to his day. *Brahmā*'s day and night put together will extend to 8640 million years. *Brahmā*'s life-span is for 100 years, each year consisting of 365 of his days and nights. These calculations will give the total duration of the universe. It is interesting to note that the Hindus never believed in the idea of a static universe which was the prevailing scientific view before Hubbell's discovery of an expanding universe. Since, according to the Hindu cosmological theory, creation, sustenance, and destruction of the universe goes on in an endless cycle, the origin of the Vedas is believed to transcend these limitations. That means their origin is beyond the concept of time and hence belongs to the realm of eternity.

With the advent of the British in India, several non-Indian scholars have researched the question of fixing a date for the Vedas from the point of their historicity. Obviously, it is difficult for a rational mind to accept the premise that even the origin of a sacred text could lie outside the framework of time altogether. It is this feeling of incredulity that provided the impetus for the early research by a group of scholars known as orientalists. This skepticism is also shared by some modern Indian scholars who have pursued a similar line of research. The researches by orientalists have resulted in different answers to the question of historicity. The absence of a definitive answer is entirely due to the questionable research methodologies that are used to arrive at the conclusions. For instance, the planetary positions which are indicated in the Vedic texts are taken as the evidence for one line of research. Yet again, another line of research has depended on the clue afforded by the style of language used. Despite the inconclusive results of the research on historicity, the orientalists have made notable contributions to the compilations of the Vedas and to their translations into several European languages. The best known scholar in this regard is Max Mueller who wrote several volumes on the Vedas with the active support of the then East India Company in Calcutta. The scholars were drawn from several countries including Great Britain, Germany, France and Russia.

Quite apart from the investigation into the question of historicity, Indians feel grateful to the western scholars for their painstaking efforts in bringing the message of the Vedas to the attention of the rest of the world. This feeling of gratitude has, however, a rider attached to it. The reservation arises from an acknowledgment of intrinsic difficulties that a scholar belonging to a particular religion faces while researching another religion. There is no suggestion of prejudice in this regard because that would be doing total injustice to the efforts at impartiality in all scholarly work. Rather, the uneasiness arises from the sensibilities most modern-day scholars have concerning inter-religious studies. It stems from the observation that a person belonging to one faith must live the other's tradition, at least for a while, in order to gain the kind of empathy that is required for its true appreciation. Faith, unlike intellectual understanding, depends upon a system of belief which is not altogether rational in character. Indians believe that the scholarly work done by most foreigners, with, of course, notable exceptions to the general statement, has not completely captured the essence of the Vedic message because it invariably belittled evidence which was not accessible to the kind of rational inquiry generally used in the acquisition of secular knowledge. This kind of skepticism is not peculiar to foreign scholars; it is also shared by many modern day Indian scholars.

Although Vedas are considered eternal, Hindus believe that major portions of the scriptures have been lost with the passage of time. In this connection, it is interesting to note the Hindu belief that there has been a gradual decadence of spiritual consciousness with the progress of the four *yugas*. It is believed that our present *Kali-yuga* marks the lowest level of spiritual consciousness, while in *Kṛta-yuga* it was at its peak. This steady decay in spiritual consciousness with the passage of time over the four *yugas* is in stark contrast to our own view of the opposite trend of the rapid rise in secular knowledge in our current *yuga*; we even talk about a knowledge explosion. It is in this context of providing spiritual knowledge for the people of *Kali-yuga* that the monumental work done by sage Veda Vyāsa is singled out for its importance. He is believed to have neatly compiled, during 5000 B.C., the Vedic hymns that were then extant into four Vedas, namely, *Ṛg Veda*, *Yajur Veda*, *Sāma Veda* and *Atharva Veda*. This was done in order to make it easy for those who are living in this *yuga* to practice the spiritual disciplines. The expectation was that, instead of trying to understand all four Vedas, one could at least try to understand the import of one Veda well. The sage is given credit of authorship to the important texts of the great epic Mahābhārata, the Gītā, the *Brahma Sūtras* and the eighteen *purāṇas*. The quintessence of the Vedic message is given in Gītā which is the popular sacred text of the Hindus. If one text has to be singled out for its pervasive influence on the Hindus, it is the Bhagavadgītā, the 'Song of the Divine'.

Brahma Sūtras constitute a concise commentary on the Vedic philosophy of the *Upaniṣads* which appear in the final sections of the Veda and hence are called *Vedānta*, end of the Veda. The principal *Upaniṣads* date back to the pre-Buddhistic period. The etymological meaning of the word *Upaniṣads* stands for 'sitting near devotedly' which is a reference to the manner of instruction given by the preceptor to the disciple. The *Brahma Sūtras* were intended to clarify all the possible misunderstandings arising from the crisp and ambiguous statements contained in the *Upaniṣads*. The three texts, namely, the Gītā, the *Brahma Sūtras* and the *Upaniṣads*, constitute the principal texts which have invited commentaries by the main exponents of the Vedic philosophy. The *purāṇas* which are meant to greatly amplify the full import of the Vedic message are of special interest. The problem of public education of Vedic knowledge was well recognized even in the historic past. Many religious scholars believe that the sacred texts of the religious literature on *Bhāgavata* were meant precisely for that purpose. The texts greatly amplify the knowledge contained in the Vedas by making careful use of myths and symbols in order to bring home the message to a larger group of people. Since the mythological stories have clear moral endings, themes of ever expanding love and devotion to God, and heroic episodes of the many incarnations (*avatāras*) of God which root out evil and uphold righteousness in this world and bless His ardent devotees with His infinite divine Grace, they combine to make the texts eminently suitable for public education.

It is difficult to imagine the scope and contents of the writings attributed to Veda Vyāsa, or Bādarāyaṇa as he is sometimes referred to. India commemorates the memory of this incredibly prodigious scholar of the Hindu scriptures by declaring a national holiday called *Gurupūrnimā* in his honor. Since Hinduism does not have a single prophet, this is perhaps the most appropriate symbolic gesture for reminding ourselves of our Vedic heritage.

Modern day Hindu sages have lamented over the fact that the number of people who can present a faithful rendering of the Vedas is rapidly dwindling. Fortunately, because of the high quality audio-visual equipment available these days, it is possible to make true copies of authentic renderings in order to preserve them for posterity. Those who have listened to a live rendering of the Vedas by a group of highly trained specialists and experienced its magic are apt to complain that the audio imprints are at best a poor substitute to the original, though they cannot deny that technology has served an extremely useful purpose.

The Vedas not only deal with matters concerning spiritual practice, but also provide guidance on the conduct of several aspects of life in a comprehensive manner. Several disciplines and sub-disciplines are spelled out in this regard within each of the four Vedas. Their characteristic feature is that each one of these disciplines and sub-disciplines when pursued to their farthest limits has the potential for helping a person towards the source of all spiritual knowledge which does not have any divisions. For example, *Āyurveda*, which is a discipline devoted to the science of medicine and the art of healing, derives its authority from its inextricable link to the primordial source of human consciousness. The discipline does not deal with just herbal medicine or naturopathy, but also is linked to the appropriate spiritual disciplines which bring about a harmony of body and mind. Similar emphasis is evident in every other discipline whether it is science, music, or literature. The exposition of each discipline should be such that it can retrace the steps from diversity to unity. This manner of compartmentalization of knowledge is, in fact, is the inextricable link between *jñāna* (knowledge) and *vijñāna* which is very often translated as science, but means more than this in the sense stated earlier. The essence of *vijñāna* is not only confined to the development of the discipline in a rational way, but it is also guided by the requirement that it should be traceable to the wholeness of all knowledge.

There are six auxiliaries to these four Vedas called *Vedāngas*. They are *Śikṣā*, which deals with euphony and pronunciation, *Vyākaraṇa*, which deals with grammar, *Chandas*, which is metre, *Nirukta*, which deals with etymology, *Jyotiṣa*, which deals with planetary motions, and *Kalpa*, which deals with procedure. In addition, there are four other texts which are also included in the list of sacred texts making a total of fourteen. These are *Mīmāṁsa*, which deals with a systematic investigation of the Vedic texts, *Nyāya*, which deals with logical reasoning, *Purāṇa*, which are mythological texts meant for greatly amplifying the truths contained in the Vedas, and, lastly the *Dharma Śāstras*, which deal with moral and ethical aspects.

Knowledge of the Vedas spread through oral transmission and as such the discipline of *Śikṣā*, the Vedic phonetics, is of prime importance to preserve tonal purity. It describes in detail how each syllable has to be pronounced. Its importance stems from the belief that even small changes in pronunciation can produce different results. The quality of the sound is of more importance than the literal meaning of the Sanskrit word in the text. Many a time, the meaning of the words are latent in the pronunciation of the words. For example, the word *daṅta* in Sanskrit means teeth. One cannot pronounce this word without the aid of teeth, as a toothless person would readily confirm. Sanskrit is a phonetic language, and it is thus important that every syllable is pronounced exactly as it is written without blurring the quality of the sound. However, some small discrepancies in pronunciation have occurred because of the influence of the regional languages of India.

Vyākaraṇa, which is the next limb of the Vedas, *Vedānga*, is due to sage Pāṇini. There are other grammars also attributed to different sages, but Pāṇini's grammar supersedes them all in importance. The text is in the form of aphorisms (*sūtras*), about which others have written detailed commentaries. It is believed that linguistics (science of language) , which is the greatest gift of man, was born at the end of the cosmic dance of Lord Siva (Lord *Natarāja*) when he sounded his percussion instrument called *dhakka* or *damaru*. There were supposed to have been fourteen beats, the same number as the total number of subdivisions of the Vedas. It is also believed that the fourteen aphorisms that were given out at the end of the cosmic dance were committed to memory by sage Pāṇini, and on that basis he wrote his text called *Aṣṭādhyāyī*, so called because it contains eight chapters. It is interesting to observe that some of the Siva temples contain the inscription of *vyākaraṇa*, and, conversely, the archeological findings in temples containing the Vedic grammar will identify Siva temples. Without knowing the context, one is apt to wonder how the dry subject of grammar is in any way connected to philosophy and religion.

Chandas refers to the metric composition of the Vedas which is an important consideration. The divisional unit of a hymn is called a *pāda*, meaning feet. The metre stipulates the number of letters in a *pāda*. It is possible for the *pādas* to be unequal in size although most commonly they will be equal. The *anuṣṭup chandas*, which has four *pādas* to a stanza and eight syllables to a *pāda*, is singled out for consideration because that is the metre used by sage *Vālmīki* for the composition of the great epic *Rāmāyaṇa*. The distinguishing feature of poetry (*kāvya*) is that there are no tonal variations as in the pronunciation of the Vedas. Metre in poetry is also important because it is mnemonic and is therefore easy to remember, unlike in ordinary prose. These technical considerations are of paramount importance because all these sacred texts were committed to memory in the absence of written texts.

Nirukta refers to etymology whereby each word is broken into its constituent syllables, and the meaning of each syllable is explained. There are several Vedic dictionaries in existence, the best known being *amara kośa*. It is believed by scholars that the subject of philology owes quite a bit to the grammar and etymology of the Sanskrit language. Of late, interest in Vedic grammar has been evinced by researchers in artificial intelligence, which is a discipline within computer sciences.

We have already referred to the Hindu cosmological theory based on the *Vedānga* of *Jyotiṣa*. This discipline was developed mainly to prescribe the auspicious times for performing Vedic rituals. It is based on the assumption that planetary motions in our solar system have definite influences on the destiny of man, and, in fact, on the world in general. The primary motivation for developing the discipline of mathematics was to render such detailed calculations possible. The Sanskrit word *ganita* means arithmetic, *avyakta ganita* means algebra, *kṣetra ganita* means geometry, and *samīkaraṇa* means equations. In India, there has been a resurrection of interest about the past achievements of Vedic mathematics. While it is legitimate to investigate the past history in order to gain a correct understanding of the country's scientific achievements, exaggerated claims in this regard will have precisely the opposite effect by calling scholarly honesty into question and thus making the effort counterproductive. It is interesting to note the assertions made by some scholars regarding several discoveries which we normally attribute to the western scientists. We read that Indians knew about the force of gravity (*apāna śakti*), which is an inference made on the basis of a statement that appears in one of the *Upaniṣads*, the *Praśnopaniṣad.* The theory of lightness, (*lāghava-gaurava nyāya*), due to the mathematician Āryabhatta, is supposed to be an allegorical reference to the rotation of planet earth round the sun. What the theory says is that it is only the lighter object that can go round the heavier one, just as a disciple goes round his spiritual master, (*guru*), who is the intellectual heavyweight. Not only this, but since it is customary in the Indian tradition to circle round the master in the clockwise direction, in an analogous fashion, it is inferred that the earth's trajectory follows a similar path because of the relative weights of the earth and the sun. While this method of inference makes interesting reading, it has, obviously, no scientific value. It might even be considered hilarious. *Bhūgola- Śāstra* means geography; the word *bhū* stands for earth and *gola* stands for sphere. The syllable *aṇda* in *Brahmāṇda* means an egg, which suggests that the fact that the world is oval was known. The credit for discovering Arabic numerals that are customarily in use is attributed to Indians. The concept of 'zero' is beyond doubt an early Indian contribution. There are several such examples which are cited in support of the intellectual dynamism that existed in the past in the scientific disciplines.

We shall now comment briefly about the sixth and the last *Vedānga* which is called *kalpa.* After a spiritual aspirant gains knowledge in the preceding five disciplines, he is considered to be ready for action. Accordingly, this limb of the Vedas spells out in detail the practical know-how of the various rituals. As this is something which takes meaning only when directly instructed by a teacher, we shall not dwell on this subject further.

We have so far mentioned the four Vedas and its six *Vedāngas.* The remaining four sub-disciplines are called *upāngas*, which mean subsidiary limbs. These are *Mīmāṁsa*, *Nyāya*, *Purāṇa* and *Dharma Śāstra.* We shall briefly comment on the contents of these disciplines.

The Vedas are broadly classified into two groups: *Karma Kāṇda*, which is an extensive treatment on various types of rituals, and *Jñāna Kāṇda*, whose focus is solely on aspects of knowledge about the Absolute Truth. Since *Karma Kāṇda* deals primarily with actions to be performed by an individual during the various events of his life to the exclusion of any mention about aspects of knowledge which is the concern of *Jñāna Kāṇda*, it is possible to draw the conclusion that there is an inherent dichotomy between the two portions of the Vedas. The insensitive abuses indulged by the priestly class over the years for their own pecuniary advantage in the observance of rituals have also lent credence to the misunderstanding. The prophets Mahāvīra and Buddha, of Jainism and Buddhism respectively, were harsh critics of rituals and so of *Karma Kāṇda* in general. Their criticisms, however, went even beyond this because they did not subscribe to the authority of the Vedas. The orthodox Hindu view, however, is that the two portions of the Vedas are meant as complementary aspects to assist the spiritual journey. The name of Sage Jaimini is associated with the first portion of *Karma Kāṇda* and is called *Pūrva Mīmāṁsa.*

The second portion is called *Uttara Mīmāṁsa*, or better known as *Vedānta*, which means the end of the Vedas. This being the case, the prefix *Pūrva* is dropped from Jaimini's thesis, and it is simply referred to as *Mīmāṁsa.* Sage Jaimini has written the aphorisms (*sūtras*) for *Mīmāṁsa*, and the detailed commentary is by Kumarila Bhatta. There is also a second commentary by Prabhākara, and accordingly, there are two philosophical schools in *Mīmāṁsa.* The principal focus of *Mīmāṁsa* is on rituals and sacrifices which are supposed to yield predetermined results. The primacy of ideas is accorded to action unlike in the *Upaniṣads* where the central thrust is towards self-realization. The following quotation from William Deadwyler [42] of the Bhaktivedānta Institute, Philadelphia, presents a good summary of the scope of *Karma Kaṇda.*

> Central to this enterprise was an extremely highly developed activity of the sort now referred to as 'ritual'–in particular, the *yajña*, or sacrifice. The Vedic *yajña* was an elaborate and painstaking endeavor, in which the learned and expert performers (*ṛtvi*), working according to the Vedic paradigm (*tantra*), had to arrange correctly the detailed paraphernalia (*pṛthak–dravya*) at precisely the proper place (*deśa*) and at the right time (*kāla*), carrying out all the prescribed procedures (*dharma*) and reciting the correct verbal formulae (*mantra*) with perfect precision. If–and only if–everything was flawlessly executed according to the most exacting standards of correctness, then the benefits for which the sacrifice was performed would accrue to the patron–the sponsor of the sacrifice (*yajamāna*).
>
> It is easy to see how the form of life that centered itself upon the Vedic *yajña* became a cult of technique, for mastery of technique was the key to power. By constructing a microcosmic image of the cosmos, and duplicating in fine the act of creation, the properly performed *yajña* gathered,

condensed, and localized the power of the cosmos itself–and so put this power into the hands of those adept at technique. Those who mastered *yajña* mastered the cosmos. The ethos of mastery through technique attained explicit expression in the writings of *Karma–Mīmāṁsa*, the philosophical school which took *yajña* as the prime Vedic *dharma*.

The principal texts of *Vedānta* are the widely known *Upaniṣads*. *Vedānta* consists of the true summary or the culmination of the teaching of the Vedas. The main concern of the *Upaniṣads* is to establish *Brahman* as the ultimate principle of the physical universe, *Ātman* as the innate sentient principle of a human being, and an identity between the two principles. The relation between *Brahman* and *Ātman* are crisply stated in various places in the *Upaniṣads* and together they are known as the *Mahāvākyas* or great sayings. The saying, *That thou art* appears in *Sāmaveda*, *I am Brahman* appears in *Yajurveda*, *Consciousness is Brahman* appears in *Ṛgveda*, and *This Ātman is Brahman* appears in Atharvaveda. However, their interpreters, who show complete unanimity of opinion in accepting the truths about *Brahman* and *Ātman*, arrive at different conclusions when it comes to the interpretation of the *mahāvākyas*.

As stated earlier, Sage *Veda Vyāsa* wrote his famous *Brahma Sūtras* in an attempt to remove all these ambiguities of interpretation. But even the *Brahma Sūtras*, for all their grandeur, are written in a very concise style consisting of 192 sections. Though small compared to Sage Jaimini's voluminous *sūtras* on *Pūrva Mīmāṁsa*, the original controversies arising from the differing interpretations of the *mahāvākyas* have remained unresolved. However, one should not lose perspective on the nature of these controversies and thereby miss the substantial amount of agreement between the various schools of thought. We will see later that there is conspicuous agreement on the practical aspects of the various doctrines arising from the commentaries. In any case, those who are treading on the spiritual path are well advised by the sages that these philosophical differences should not serve as deterrents for spiritual practice as such. Despite this sagely advice, the sectarian differences persist between the followers of various philosophical schools to the extent of stifling the real purpose of the scriptures.

The metaphysics of the various *Upaniṣadic* traditions vary widely in their views about the relationship of *Karma kāṇda* (*Mīmāṁsa*) to *Jñāna Kāṇda* (*Upaniṣads*). Broadly speaking, there are three views that can be discerned. First is the declaration of absolute supremacy of *Karma Kāṇda* over *Jñāna Kāṇda*; second, we have the exact opposite view declaring the supremacy of *Jñāna Kāṇda* over *Karma Kāṇda*; and third, the view that the two are complementary in nature. The first view which is held by the protagonists of the *Pūrva Mīmāṁsa* school logically leads to the denial of the existence of God or His irrelevance to philosophical inquiry. The second view categorically refutes the claims of the *Mīmāṁsa* school and establishes the primacy of *Brahman* as the central principle of cosmic unity or, put another way, it establishes the supremacy of God. The complementary view is the more moderate one, and it is perhaps the correct view when the distortion supporting the *Karma kāṇda* is avoided. The idea that *Mīmāṁsa* by itself has an autonomy of ideas quite independent of the message of *Upaniṣads* has to be abandoned. It should be viewed as a preparatory stage for entering into the study and practice of the *Upaniṣadic* message. In this connection the intensive debate on the controversial question by sage Śaṁkarācārya with Maṇdana Misra, who was champion of *Mīmāṁsa* is often cited in the literature. It was only after Śaṁkarācārya convinced Misra about the comprehensive nature of the *Upaniṣadic* message that Misra became a disciple of the former.

The next *upānga* is called *Nyāya* which is the science of reasoning and is attributed to sage Gautama. The discipline arose in the context of establishing that *Īśvara* is the creator of the universe. The discipline gives a detailed account of all the rational methods of proof pertaining to the physical world and also considers the Vedas as the testimony for philosophical reasoning. It is interesting to note that *Nyāya Śāstra*, the logical foundations for reasoning, was a well-established discipline in India, although scientists are accustomed to giving exclusive credit for their origins to ancient Greece. One is reminded of Aldous Huxley's refrain about a 'spiritual iron curtain' that was drawn in ancient Greece which made it impossible for a free exchange of ideas between the East and the West. It is only of late that this curtain is slowly coming up resulting in an enrichment of both Eastern and Western thoughts.

We commented earlier on the importance of *purāṇas* in connection with the colossal contributions made by sage Veda Vyāsa. Vedic aphorisms lend themselves to interpretation in the form of stories making artful use of myths and symbols. For instance, the aphorism *satyam vada* which is the moral injunction to speak the truth is the subject matter of the familiar story of king Hariścandra. There are eighteen *purāṇas* which deal with several aspects of the Vedic message. Vedic scholars claim historicity of the stories that are told, but irrespective of the veracity of this claim, no one can deny the sheer grandeur of this style of popularizing ideas connected with ethics and the essence of spiritual knowledge.

The fourth and the last *upānga* is *Dharma Śāstra* which gives a detailed set of moral injunctions combined with actual procedures that a spiritual aspirant should follow during the course of his life. Thus, the discipline deals with the subject of ethics, not as an end in itself but as a means for advancing on the path of spiritual fulfillment. It is only in very simple situations that one can discern cause and effect relationships governing one's actions and decide on what is good and bad for one. As the situation becomes more and more complex, it becomes increasingly difficult to decipher the cause and effect relationships, and consequently, the task of separating good from bad becomes formidable. The modern-day world provides plenty of examples of acute conflict of this type in all spheres of life. Furthermore, since moral and ethical cleansing is considered absolutely essential for making spiritual progress, detailed guidelines become necessary as to how one should mould one's course of life in order to achieve these aims. In all such cases of both secular and religious matters, one needs some guidance for following the path of right action. Philosophically speaking, we can say that the overall constraints should be such as to ensure that one's actions are always in consonance with the cosmic law governing the universe. But this intellectual understanding cannot provide a moral compass because we do not know exactly what that cosmic law is, beyond a recognition of its possible existence. It is here that the *Dharma Śāstras* come to our rescue. They are formulated by sages who have had experience of the ultimate truth, which is what gives the discipline its veracity. The *Dharma Śāstras* are very extensive in scope embracing aspects of both secular and religious life.

In addition to the above list of fourteen, we may also include four *upaṇgas* which are appendices to the main vedic texts. These are *Āyurveda*, the science of life, *Artha Śāstra*, which in modern terminology is called Economics, *Dhanurveda*, which deals with weaponry, and *Gandharva veda*, which includes all the fine arts such as music, drama, and dance.

There are six systems of Indian philosophy which are usually presented in three pairs: *Nyāya–Vaiśeṣika*, *Saṅkhya–Yoga*, and *Pūrva Mīmāṁsa– Uttara Mīmāṁsa* (*Vedānta*). The pairing is due to their proximity in their philosophical tenets. Nyāya is associated with Sage Gautama, *Vaiśeṣika* with Sage Kanada, *Saṇkhya* with Sage Kapila, *Yoga* with Sage Patanjali, *Mīmāṁsa* with Sage Jaimini and *Vedānta* with Sage Vyāsa. It is possible to view these individual philosophical schools as different strands of a coherent whole in terms of modern thinking on systems theory. This statement will be better understood when we have discussed some of the individual philosophies in some detail. In this text, we will concentrate on the principles underlying *Saṇkhya*, *Yoga* and *Vedānta*. An exposition of all the Vedic schools of philosophy is beyond the scope of our outline.

Sanskrit is the liturgical language of Hinduism. Many of the modern languages of India are its offshoots. The script itself is derived from the original *Brāhmī* script which, in turn, is the parent of the scripts of the various regional languages of the country. Sanskrit, which is also called *Devanāgarī*, has 52 letters in the alphabet and is a phonetic language encompassing all sounds that a human being is capable of producing. There are good word processors which are available for almost all of the scripts.

There is a common misconception amongst Hindus that without a knowledge of Sanskrit it is impossible even to take the preliminary steps for the understanding of Indian philosophy. Unfortunately, this myth is wittingly or unwittingly perpetuated by many scholars, with the result that many just give up their philosophical pursuits because of this perceived handicap. While it is true that for a scholarly understanding of the texts, one must have a thorough grounding in the language in which they are written, the main kernel of the message can, however, be successfully transmitted in any well-developed language. Apart from the truly outstanding scholars, the number of non-Indians who have made great headway in understanding the Vedic message bears testimony to the assertion. The purpose of drawing attention to this fact is to not let the lack of knowledge of Sanskrit act as a deterrence to those interested in Hindu philosophy.

2.2 The Essence Of Vedic Philosophy

Of all the various sections of the Vedas, it is the *Upaniṣads* which has caught the attention of the modern mind of both the East and the West because they go to the very heart of the Vedic message with all its universal appeal. The *Upaniṣads* come at the very end of the systematic compilation of the voluminous texts, and hence they are also called *Vedānta* which means the end of the Vedas. Since they provide the basic conceptual framework for the declaration of the ultimate truth and its subsequent discussion, they form the most important passages of the Vedas. Some of the tersely stated key ideas of the *Upaniṣads*, the *mahāvākyas*, pertain to the cosmic principle governing the universe on one hand and the innermost essence of Man on the other, as well as the interrelationship between the two apparently dual concepts. We can restate the same idea in terms of the operations of the mind that we have developed in the last chapter. The cosmic principle refers to the root of the universal mind, and the innermost essence of man refers to the substratum of the individual mind.

While there is no controversy at all within the Vedic schools about the ultimate truths underlying Man and the universe, their interrelationship is amenable to more than one interpretation; therefore every established teacher of the past has made it a point to offer clarifications in this regard for the affirmation of his own insights. These differences in philosophical interpretations do not masquerade the fact that there is a substantial amount of commonality amongst them.

What we have called the Vedic philosophy is the interpretation based on *advaita*, which is the nondualistic interpretation due to the great sage Śaṁkarācārya, who lived about fifteen centuries ago. While recognizing that there are other interpretations of great merit that are current even to this day, it cannot be denied that the nondualistic interpretation is the one that has received wide publicity and acceptance. It is also the one that has enamored most scientists who are interested in Hindu philosophy and religion. This view is strengthened by the fact that there is a great deal of concordance between philosophical insights arising from *advaitic* study and those stemming from the various paradigms of science. While making this observation, it is pertinent to remind ourselves again that science has long ago given up the goal of unveiling the ultimate truth behind the universe. In fact, it is now clearly recognized that metaphysical truths lie beyond the reach of scientific rationality. But the limits of scientific thought do take us to the no man's land between science and philosophy, and this grey area is accessible to both philosophical and scientific reasoning. The topics of free will, anthropic principle, argument from design, theory of evolution, and nature of diversity of the universe are some of the cases in point, all of which provide deep insights into the metaphysical problem.

Our choice of starting out from a discussion of nondualistic interpretation is not motivated by any rigid preference for the *advaitic* school of philosophy. We begin with a very brief survey of ideas pertaining to nondualism only as an excellent point of entry to the study of the vast literature. The principal teaching of nondualism focuses attention on the ultimate realities in the realms of matter and spirit and arrives at the irrevocable conclusion that there is one and only one ultimate reality which is spiritual in nature. This grand conclusion was greatly assisted by the long history in Indian philosophical thought which established a parallelism between the study of the individual and that of the universe as a whole. This was an attempt to understand and express the ultimate reality of the universe in terms of its counterpart at the level of the individual. The eternal principle as realized in the world is called *Brahman.* It can be construed as the *Cosmic Principle* whose manifestation is the universe we are living in. *Brahman* is the *transcendent Absolute.*

It is important to note that the discovery of the Cosmic Principle is not arrived at by successfully establishing a grand unification of all the laws of nature, as one would attempt to accomplish through the discipline of physics. This principle is beyond the reach of physical theories of Nature such as Newtonian mechanics, relativity theory and quantum theory, or the theories of life sciences such as Darwin's theory of evolution. Nor is it dependent on the cosmological investigation of the origin of the universe as exemplified in the big bang theory. While this is so, one might legitimately question the validity of existence of *Brahman* as a cosmic principle which transcends the realms of the farthest reaches of rational inquiry of the scientific method. It is true that when approached from this point of view, the principle remains a mental construct, an elusive concept at best, without any possibility of scientific verification. It should be understood, however, that when we say that it is beyond scientific inquiry, we are not meaning thereby that it is unscientific, only it belongs to the realm of metascience. Whether we give credence to any concept which is not within reach of reason is a matter of taste and preference. That is why there are scientists who subscribe to religion and those who do not.

The validation for existence of the cosmic principle is sought by taking an entirely different approach. In the last chapter, we said that the paradoxical union of being and becoming is definitely a human experience beyond a shadow of doubt and that we implicitly assume the existence of a similar phenomenon in the universe at large. The universal mind therefore is only a mental construct based on plausible arguments. The hypothesis for the existence of such a cosmic phenomenon is based on our intellectual understanding, based on reasoning. Specifically, unlike the individual mind, it is not based on actual experience and consequently, it is not at all self-evident. Appealing to formal methods of logical reasoning and drawing the right conclusions for a correct understanding is the only recourse we have for this study. We find ourselves in the rather awkward situation of having to invest the feature of reality to the universe at the very outset and then to start investigating what this reality is all about through logic which is itself an ingredient of the universe we are investigating. There is no other alternative to this route if the goal is to investigate the reality of the universal mind. Scientists have followed this line of research, and so the development of western philosophy is inextricably linked to natural philosophy.

Hindu philosophy, on the other hand, has preferred to probe into the mystery of this universe starting from what is experienced at the human level. This takes us on to an inward journey to the innermost recesses of the human mind rather than on a Star Trek exploration into an investigation of the mystery of our external universe. For purposes of our further discussion, we shall tentatively assume that a withdrawal from sensory experience even while remaining in the conscious state is entirely possible in order to facilitate an inward journey. Instead of trying to identify the core truth behind the observed phenomena, which, in any case, is beyond the pale of our human experience, we follow the alternative course of investigating what constitutes the innermost essence of man and the observer in him. This is also referred to as the *Psychic Principle* whose exploration at every stage is inextricably linked to one's own preconditioning of the mind. The word psychic in this context does not refer to anything even remotely connected to the modern discipline of psychology; rather, it is related to the traditional meaning of Indian psychology that refers to the science of the soul. Every sensory experience leaves a trace in one's memory bank, and it is the cumulative effect of this is what we call preconditioning. Obviously, the precondition of one's mind is different from another's because of the variegated nature of experiences and the manner in which they imprint on one's memory due to the special way we look at things and absorb the impressions that are recorded by our mind. The psychic principle manifests itself only when we are able to withdraw from our sensual experiences altogether. This is no ordinary feat, but mystics throughout the ages belonging to every religion inform us that it is entirely possible to achieve that sublime state. Furthermore, they assure us that it is, in fact, the most natural state of a human being.

Fortunate are those who have had this experience without prolonged effort, but for the majority of mankind, the strict observance of some spiritual disciplines become mandatory in order to make any progress at all in the right direction. The realized souls proclaim that the psychic principle, which is deep-seated within us, has a sense of undeniable reality attached to it in the same sense that one does not look for a proof for reality of one's own existence. It is the real subject of all experiences, the eternal witness (*sākṣi*). More than that, it is a state of mind where one does not experience the paradoxical conjunction of being and becoming. The mystics, belonging to different countries, religions and historic periods, who have realized the truth of the psychic principle have a common story to tell the rest of the world, and it is a compelling story whose authenticity is difficult to dismiss on the basis of our common understanding of subjective experience with all the uncertainties associated with it.

It is important to reiterate that the so-called subjective experience of the psychic principle has to be clearly distinguished from other subjective experiences that we normally refer to. The former is an intuitive experience which arises only when the senses are withdrawn unlike the latter in which they have to be very much present. It is because of this clear-cut distinction that the experience of psychic principle cannot be pigeonholed into the category of subjective–objective experiences of our worldly realities. Moreover, it should not be overlooked that the possibility of realizing the existence of the psychic principle usually comes after long years of preparation. It is not an instant experience accessible to everyone, a feature it has in common with experimentation in advanced science which also demands long years of preparation. There are, of course, reported instances of a very rare breed of people who have gained self-realization very quickly, but these have to be treated as exceptions only. Those who are born with self-realization are called the *avatāras*, and they descend to this earth for the benefit of mankind. But in the majority of cases, spiritual experience can only be gained through some time and effort. The exceptions to this rule have to be treated in a separate category altogether.

We have talked about two principles: first, the cosmic principle called *Brahman* which is the principle forming the basis of the universe as a whole and second, the psychic principle which is the substratum of the individual self. This psychic principle is called *ātman*, and the unique declaration of the *Upaniṣads* is that the cosmic and psychic principles, the *Brahman* and *Ātman*, are one and the same. There is a complete identity of the two principles. We quote from M.Hiriyanna's book on *Outlines of Indian Philosophy* [21]:

> Thus two independent currents of thought–one resulting from the desire to understand the true nature of man and the other, that of the objective world– became blended and the blending led at once to the discovery of unity for which there had been a prolonged search. The physical world, which according to the *ātman* doctrine is only the not-self, now becomes reducible to the self. The fusing of two such outwardly different but inwardly similar conceptions into one is the chief point of *Upaniṣadic* teaching and is expressed in the 'great sayings' (*mahāvākyas*) like 'That thou art', ' I am *Brahman*' or by the equation *Brahman* = *Ātman*. The individual as well as the world is the manifestation of the same Reality and both are therefore at bottom one. There is, in other words, no break between nature and man or between either of them and God'.

The identity of *Brahman* and *Ātman* is of special significance. The concept of *Brahman* as the cosmic principle bereft of this identity will forever remain in the realm of speculation since there is absolutely no validation for it either in terms of our investigation of Nature or in terms of our personal experience of it. On the other hand, when its identity with Ātman is recognized, it immediately becomes a *spiritual principle*. The route to the substratum of the universal mind is by recognizing its identity with the substratum of the individual mind. Since the latter is based on personal experience, albeit of a trained mind, there is no uncertainty attached to it, and in view of the identity, the cosmic principle is also divested of all uncertainty. We had earlier stated that the universe is a manifestation of the cosmic principle. In view of its identity with the psychic principle, we come to the startling conclusion that the universe is indeed a manifestation of a spiritual principle. The corollaries which flow from this new understanding have profound implications. For instance, it implies that spiritual evolution supersedes biological evolution, which is a proposition that can be discussed at length on its own merits. In this context, biological evolution means not only the appearance of life but also of consciousness and self-awareness. This conclusion about the nature of the cosmic principle does not deny the validity of biological evolution; only it asserts that it is the spiritual element that is its ultimate causative factor.

One can now examine the second important implication of the identity relationship. Earlier we referred to *ātman* as the psychic principle in order to distinguish it from the cosmic principle of the cosmos as a whole. The psychic principle, since it is experienced by the individual self, can easily be mistaken for something which is characterized by diminutive dimensions of finititude. But its identity with the cosmic principle removes this doubt once and for all. It proclaims that the psychic principle is also infinite in character. One can see that it is the symmetry between *Brahman* and *Ātman*, and their identity, that removes the uncertainty associated with the former as an isolated concept, and at the same time invests the latter with the expanse of infinity. While mankind has been described as animated dust when compared to the dimensions of the universe, we can nevertheless entertain the robust optimism that the inner essence of Man enjoys the same status of the cosmic principle governing the universe. This is, undoubtedly, the most unique contribution of the *Upaniṣads* and constitutes the core of the Vedic philosophy.

There is a beautiful Sanskrit word which sings the glory of *Brahman* and *Ātman* and their identity. It is called *saccidānanda* and consists of three epithets, namely being (*sat*), sentience (*cit*) and bliss, (*ānanda*). These three can be visualized as the three legs of a tripod in order to portray the resulting stability of the mind with all its superlative intrinsic qualities. The qualities implied by the three epithets should therefore not be mistaken for qualities that the seeker attributes to that state in order to extol it. Rather, they are natural qualities of that state. Being is existence, and that by itself will not guarantee the spiritual dimension which is characteristic of sentience. Furthermore, it is the element of total bliss which invests the first two with a unitary character.

In our day-to-day lives, we find by careful observation, occasions when we experience one or the other of the three intrinsic qualities of existence, sentience and bliss. For example, the blissful experience after getting a good night's rest is an universal experience. It is such experiences, which the poet Wordsworth called the 'intimations of immortality', which can be traced to their ultimate source of *Brahman* or *Ātman*. One is reminded of an analogy arising out of the Big Bang theory of the universe in this regard. We are referring to an indirect evidence for the historicity of creation of the universe. The relatively recent detection of micro-wave radiation enveloping the entire universe, which was confirmed through experiment, lends support to the theoretical surmise of the creation of the universe through an infinite explosion some fifteen billion years ago. Similarly on the human scale, there are many clues during our worldly experience, if only we carefully observe them, which are strongly suggestive of the *saccidānanda* character of transcendence of the cosmic principle and of the immanence of the psychic principle, of *Brahman* and *Ātman*.

We can now examine the full import of this transcendental reality of *Brahman* to the physical world that we live in and to our selves. We know that one of the essential characteristics of our world is diversity. There is an endless proliferation of material objects surrounding us, a multiplicity of individuals, knowledge which is exploding at a fast pace and the universe itself which is expanding. Knowledge of this empirical reality of the phenomenal world in all its multiple facets is the 'lower truth', *aparā vidyā*. It is called 'lower truth' as viewed from the platform of the Absolute, the *Brahman*, which is the sole reality with its unitary character. This latter truth is called the 'higher truth', *parā vidyā*. This classification of knowledge is given in the *Mundaka Upaniṣad*. While experiencing the 'lower truth', which we call worldly reality, we entertain no doubts whatsoever about the specificity associated with the diversity of our universe; we unconsciously subscribe to the fragmented view that one object is essentially different from the other. Each object of the universe is identified by a distinct label, a name (*nāma*) and form (*rūpa*) which together gives it the specificity; further, when the world manifests from its undifferentiated source of *Brahman*, the differentiation is in terms of *nāma* and *rūpa*. But these views, born out of the feeling of our separability from the rest of the universe, that is the feeling of duality at the empirical level, will vanish once the higher truth is realized.

The feeling of separability is a direct consequence of the physical aspects of the body and mind, whereas the true Self, which is the *Ātman*, is always inseparable because of its identity with *Brahman*. These physical aspects are called limiting adjuncts, (*upādhis*), the constraints operating on the Self, which produce the limiting feeling of finititude. They are limiting constraints only in the absence of an actual realization of the psychic principle. Mere intellectual recognition of the problem can not remove these barriers. From the vantage point of worldly consciousness, the constraints are experienced as very real. In other words, the same phenomenon is experienced very differently from the two planes of transcendental and empirical existence. Based on these differing perceptions, we can safely conclude that it is the reality experienced in transcendental existence that is the independent reality since it is never sublated, while the reality experienced in empirical existence assumes the status of a dependent reality. This metaphysical truth, however, remains veiled in our day-to-day worldly experience. This obscurity is attributed to spiritual ignorance (*avidyā*), which is deeply ingrained in all our habits of thought and modes of action. The cosmic counterpart of *avidyā* is called *māyā* which is the concealing power that totally screens out the ultimate reality of the universe.

The primary purpose of spiritual life is to put an end to *avidyā* so that one can gain spiritual enlightenment which is called *jñāna*. When this goal is achieved through spiritual practice, one attains the true state of enlightenment which is called *mokṣa*. The theoretical aspect which results in intellectual comprehension of the goal is the philosophical aspect, while the practical aspect of spiritual practice is the Vedic religion. The two aspects, philosophy and religion, which come under the guidance of the Vedas are inextricably linked. One without the other would be incomplete. In fact, this is what brings Vedic philosophy closer to religion.

The practical aspect would first of all need the guidance of a spiritual master, a *guru*, who has already realized the ultimate truth. This requirement might be difficult to fulfill because of the rarity of this extraordinary breed of people, but its theoretical import should not be missed. Moreover, an in-depth discussion of the concept of *guru* will point to other possible avenues that will spontaneously open up to a spiritual aspirant, the most important of which is the guide from within.

Spiritual practice devised for overcoming spiritual ignorance consists broadly of three stages. The first is hearing, *śravaṇa*, the second is contemplation, *manana*, and the third is meditation, *nididhyāsana*. More will be said about these later. For the present, it is sufficient to mention that the *śravaṇa* stage is meant for hearing about the philosophical truth. Since the message is rather abstract, the profundity of the theme needs constant repetition in more than one way. When the message has made its total imprint, one proceeds to the intellectual phase of *manana* where one reflects on the truth of the new learning. The emphasis here is that the aspirant should not blindly believe whatever the preceptor (*guru*) has taught, but should subject it to a critical examination before finally accepting its possible veracity. Belief systems, as we know, range all the way from blind faith and indoctrination to well-founded reasoning based on logical principles. What is recommended here is a non-dogmatic attitude based on critical thinking. Since the existence of the Cosmic and Psychic principles, and their identity, is something that can be accepted only when there is at least a fleeting experience of it, the preparatory intellectual phase of understanding is invariably shrouded with uncertainty.

At this stage of constant contemplation, one can only assign a high degree of probability to the metaphysical truth on the basis of collective intellectual evidence, but nothing beyond that. In order to remove the lingering uncertainty, one enters the third and final stage that is called *nididhyāsana* where the final goal is experience and not mere intellectual understanding. This is the practical phase with multiple aspects to it, but the primary aspect is meditation on the truth that has been comprehended at an intellectual level, albeit with all its uncertainties. It is the phase where one subjects oneself to a rigorous training in order to realize the truth through intuition.

The emphasis on the allied spiritual disciplines for achieving the goal is very important in this context, and needs reiteration because very often we see exclusive concentration on meditation only. The faculty of intuition is also a particular mode of the mind, and it can be construed as the very antipode of the mental faculty responsible for rational thinking on the basis of sensory data. The practical training has several facets to it and is best understood when one actually gets started on this phase. No amount of theorizing about it will give the precise idea. What one can say emphatically about it is that it calls for a rigorous discipline in order to be able to achieve moral and ethical cleansing. The traditional wisdom on this phase of activity is that, in fact, the cleansing process should precede before one launches on a discipline of meditation. Practically speaking, however, the motivation for the purification process comes only when one has at least a momentary experience of the benefits of meditation. The analogy that one can provide for a sharper understanding of the process of the mutual interaction of meditation and the practice of cleansing is to describe it as a *feedback process.* Moral and ethical cleansing will assist meditation which, in turn, will strengthen our ability to engage in the cleansing process. But, a view expressed by many spiritual teachers is that one should commence meditative practice first in order to derive the requisite motivation for moral and ethical cleansing.

Karma mukti is a concept which stands for gradual realization. The word gradual should be understood in a special sense in view of the fact that the goal of the spiritual efforts is right within us. It is not a journey that we undertake to reach a destination that is distant from us. The concept of distance is not as we understand from a road map. It is the kind of distance which is very near and yet very far, a 'non-metric distance' as we say in mathematics. Progress through stages is achieved in a gradual manner because it does take time and effort to ward off the effects of the constraints, the limiting adjuncts, which are operating on the *ātman*. These conditions in a specific instance are determined by the nature of one's own past deeds and thoughts. By past, we mean not only the past in one's present life but also in one's past lives. There is no testimony other than what is provided by the scriptures for the belief in reincarnation, although scientists have also been dabbling in exploring this phenomenon.

The question of incarnation has reference to the *Karma doctrine* which will be discussed later. At this stage, it is enough to know that the doctrine assures us a system of rewards and punishments based on an unerring principle of moral retribution. This is an inexorable deterministic principle akin to Newton's third law of motion which states that action and reaction are equal and opposite. Good deeds and thoughts will always lead to a good end, and the converse is also equally true. It now remains to explain what we mean by good and bad deeds or thoughts. When we say that the doctrine is deterministic, we mean that there is a cause and effect relationship. It is possible to arrive at some understanding of what is good and bad when the causal relationships are simple and based on established ethical conduct applicable to such situations. Ethical dilemmas, however, become progressively acute when the causal relationships become increasingly fuzzy, as in complex situations. In fact, the acuteness of such dilemmas increases in proportion to the moral virtuosity of the individual experiencing it because of his heightened sensitivity to the consequences of the conflicting issues. The dilemma of Arjuna on the battlefield of *Kurukṣetra* as narrated in Gītā is a case in point. These are situations, where what is the right course of action cannot be decided by being anchored to the intellectual plane of existence where rationality reins supreme. In Hindu philosophy, the subject of ethics is given a prominent place, but good ethical behavior is viewed as something that naturally follows from the relentless pursuit of realization of the psychic principle. Ethics when treated without reference to the psychic principle is stripped of all its real value for continued sustenance.

We have so far given a brief account of Vedic philosophy, but there has thus far been a conspicuous omission of any mention of God. But the concept of God is absolutely essential for the practice of the Vedic religion as indeed for any religion. The version of Vedic philosophy that we have discussed is called the absolutistic version of *Vedānta* which is often criticized for its impersonal bearing. It is considered by many as a hopelessly abstract topic to be of any practical value. However, it will remain impersonal only until such time that the conception of God consistent with the twin concepts of *Brahman* and *Ātman* and their identity is not recognized. The absolutistic version highlights the aspect of transcendence, while the theistic component emphasizes the aspect of immanence of God. He is considered to be immanent in all the constituents of nature. In fact, the whole universe is viewed as His cosmic vesture. God, in this nondualistic version of *Vedānta*, is the personified version of *Brahman*, which is the cosmic principle with all its infinite capabilities. The concept of the Vedic God is not akin to that of the Semitic religions where He is portrayed as standing above man and nature thus leaving out His aspect of immanence.

So it is possible within this framework of Vedic philosophy to find the appropriate place for the concept of God which is so essential for the corresponding religious practice. We are assured of his lofty place with a definite axis towards the devotee. On the other hand, He is considered to be both the creator, as well as the very stuff that creation is made of. To put it more succinctly, God is considered to be both the effective as well as the material cause of the universe analogous to a spider weaving a web from the material it produces from within.

We will discuss the message of this chapter in greater detail as we proceed. In preparation, we shall introduce, in the next chapter, some additional concepts and terminologies which frequently appear in all discussions on Hindu philosophy. We have also included some scientific concepts which illuminate the grey area between science and Vedic philosophy.

Chapter 3

Some Basic Concepts of Vedic Philosophy

We now discuss an assortment of philosophical and scientific concepts which are meant to throw further light on the interpretation of the central message of Vedic philosophy. Some of these concepts have already appeared in concise form in our earlier discussions, but we deem it useful to dwell on them further in more detail. In fact, some of the topics discussed here are of general interest reaching beyond any particular school of philosophy. The order in which the subject matter is presented does not indicate any preference for the order of importance of the topics.

3.1 A Conception of Values

The perennial question of determining the ultimate truths hidden behind man and nature has been answered by diverse philosophical systems. The very fact that no single system has gained universal approval points to the fact that intellectual speculations on the mystery of the universe admit of a diversity of solutions. Notwithstanding this lack of clear answers, however, it is interesting to observe that philosophical investigation is not considered a fruitless activity. On the contrary, it continues to occupy our time and effort based on the simple fact that philosophizing comes very natural to man. Any deep analysis of the general human condition results in a discussion of philosophical premises which account for it. Also, the pursuit of any discipline to its logical limit culminates in a philosophical investigation of its moorings. It is abundantly clear that man feels an insatiable urge to formulate some conceptual framework to understand the true meaning of any new important factual knowledge he acquires . He also recognizes that in the absence of such a mental framework, experiences based exclusively on sensual perception could be misleading. This basic conviction gives him the tremendous urge to strive for an intellectual vantage point from which he can decipher the significance of the myriad experiences that he has had in life in a comprehensive way.

In the early stages of human civilization, it is said that man was seized with wonder and curiosity about the world he was living in. This observation suggests that he was not completely devoid of all knowledge of himself and his environment, but already possessed a rudimentary knowledge of it. Both wonder and curiosity are characteristics of partial knowledge, but not of a complete lack of it. For instance, if one knows nothing at all about the Big Bang Theory explaining the origin of the universe, one will not experience any wonder or curiosity about this important cosmic event. If one possesses only an elementary knowledge of it, one would expect the necessary incentive on his part to know more about it. It is partial knowledge that gives the impetus to gain a fuller comprehension whereas the total absence of any knowledge will not be accompanied by wonder and curiosity. In the language of modern day probability theory, which is a branch of mathematics, it is only from a knowledge of the sample can one hope to devise methods of inference about knowledge of the entire population. The traits of wonder and curiosity can never be fully absent because at any point in the evolution of human civilization, man's knowledge about the universe will always remain partial since the diversity of Nature is constantly on the increase. The expansion of the universe at any given moment should lend support to this observation. Furthermore, the final conclusion is not limited to the physics of the universe, but is true of all types of worldly knowledge.

The aim of philosophy is to know the whole of reality, not just fragments of it as in the case of scientific knowledge, or other aspects of secular knowledge. For example, the disciplines of physics, chemistry, biology, psychology, economics, etc., deal with specialized branches of knowledge, whereas philosophy is interested in knowing the ultimate truth behind the universe. Furthermore, the sum total of all the separate investigations of the various disciplines will not amount to the inquiry that philosophy is concerned with. The reason for this conclusion is that the sum of all the fragments of knowledge will not total up to the whole which is a proposition that can be proved with rigor.

The ignorance arising from partial knowledge in search of philosophical truth is of a special variety called spiritual ignorance. In Sanskrit, it is called *avidya.* It has to be distinguished from the kind of ignorance associated with partial knowledge of our everyday experience whether it pertains to the external world or to our own individual lives; this latter kind of ignorance can only be removed by new knowledge that is acquired through effort. The steady proliferation of intellectual disciplines bears testimony to this observation.

The unique nature of spiritual ignorance, on the other hand, can be understood by means of an analogy. It can be compared to a cloud enveloping the bright Sun; the Sun is ever present, and all that is required to see the Sun is to devise a means for clearing the cloud. The Sun is analogous to ultimate reality, which is already there; no new knowledge is required to establish its existence, and, in theory, it is something that is acquired effortlessly. Spiritual ignorance is analogous to the cloud of unknowing, and the knowledge required is meant to dispel this cloud in order to apprehend the truth; acquisition of this knowledge does not ask for immediate access to a well-stocked library or a good search engine on the world wide web. It is the critical examination of the consequences of our self-awareness of our spiritual ignorance that constitutes the initial point for the inquiry. It is the kind of analysis, based on experience and reflection, coupled with guidance from those who have already traversed this path, which provides the road map for determining the way out of spiritual ignorance. The importance of self-awareness for philosophical inquiry is highlighted by the fact that animals do not indulge in philosophical speculations because they lack the faculty of self-awareness, which is unique to homo-sapiens.

As said earlier, the nature of spiritual knowledge is clearly distinguishable from that of scientific knowledge. The methodologies employed for their acquisition are quite distinct. In order to acquire new scientific knowledge, we first form a hypothesis and investigate a phenomenon to come up with new facts. This investigation can take on several forms: a thought experiment, a theoretical study, a computer simulation, or a laboratory experiment. When seemingly isolated new facts are observed, we attempt to fit them into a coherent theory and thus gain a new intellectual understanding of the phenomenon under investigation. However, we do not insist that these facts have some bearing on the conduct of our lives, either directly or indirectly. Nobody expects Einstein's theory of relativity, for instance, to have practical applications to our lives. In other words, there is a general consensus that science, by its very nature, is neutral with respect to values.

Unlike scientific knowledge, philosophical knowledge will have relevance only in the context of values. We insist that philosophy must have some practical applications to life. If philosophical knowledge is divested of this, then it will resemble scientific knowledge, being relegated to the purely intellectual realm. Indian philosophy lays great emphasis on the difference between facts and values and goes to the extent of proclaiming itself as primarily a study in the Indian conception of values. One incidental advantage of focussing on values is that philosophy lends itself to an empirical bottom-up approach of development, evolving systematically from the basic building blocks of human experience. We proceed in this fashion until the final stage, at which the formulation of a philosophical ideal becomes absolutely essential.

In the beginning of human civilization, man's primary motivation was self-preservation in a hostile environment. Subsequently, he felt the desperate need to gain at least some empirical knowledge of the environment in order to cope with the day-to-day aspects of his practical life. It is well known that an imperfect understanding of the physical surroundings can seriously compromise our safety. It is difficult to lead purposeful lives under inhospitable conditions, and, consequently, there was an enormous urge to free ourselves from the tyranny of such constraints. The progress of civilization has been marked by the progressive mitigation of this *physical evil*, which man is confronted with. What started out as a haphazard acquisition of elementary knowledge of the environment without any thought given to assimilating it within a conceptual framework has now gradually evolved into scientific knowledge with all its rigor and versatility. We have well-proven theories about nature, which have very wide implications because of their capacity for generality and predictive value. On the basis of these scientific theories, a wide range of technologies has been developed with the capability of eradicating the physical evil to a great extent. It is now possible for mankind to live in comfort if only there is the will and wisdom to use the sophisticated technologies to that end.

However, there is clearly a flip side to the spectacular achievements of modern-day technologies. The pendulum of technological progress has swung too far in the direction of gaining control over nature, thereby abandoning man's early emphasis on combating the physical evil without disturbing his harmony with nature. The price we pay for this unintended drift in technological progress is the destruction of the environment with all its adverse long-term consequences for mankind. The motivation for the new economic man is no longer limited to ensuring self-preservation; instead, it is to appease the insatiable desire to acquire more and more pleasures and security in life. Animals also seek the satisfaction of their desires for pleasure and security in life, but they are propelled by instinct and not by conscious choices as in the case of man. Choice can be exercised in the direction of seeking more gratification than there is an actual need for. In animals, on the contrary, the impulses are naturally disciplined by the governing instincts.

The limitless and undisciplined impulses that are unleashed in the process of satisfying man's desire have the potential to create havoc for mankind. They have resulted in a situation where the physical evil, rather than being mitigated, has been aggravated because of the mindless use of technology aimed at increasing the material well-being of man. The cost involves irreparable damage to the environment and gross injustices to fellow human beings. Only recently has the knowledge dawned on us that the global environment has to be protected; as a result, we are becoming bitterly resigned to the stark fact that there are natural limits to the acquisition of pleasures and security. Technology does not address the question of values, but neither is it indifferent to them. It can be a boon, when used wisely, for not only overcoming the physical evil, but also for creating positively comfortable circumstances for all people without threatening the environment.

The pursuit of selfish ends gives rise to the *moral evil*, and the knowledge for its removal is the province of philosophical inquiry. Moral evil has its roots in human nature, and it has existed since the dawn of human civilization since it is related to man's ability to choose his actions. Only, this evil appears in an exaggerated form in our technological age because the stakes are now much higher. The starting point for the study to come to grips with this problem is again man's remarkable capacity for self-awareness. It is essential to gain the necessary philosophical knowledge to usher in ever-increasing degrees of enlightenment so that man can make the right choices, which are conducive to his own ultimate good as well as to the community at large. We should put great effort into preserving the human race rather than follow the reckless path of exercising unregulated control over nature. It is only by pursuing such aims that man can continue to live in harmony with nature after successfully acquiring the scientific and technological knowledge to overcome the physical evil.

Moral evil can only be confronted by the right type of philosophical knowledge, which, in turn, means that we must clearly formulate the ultimate end towards which an individual should strive in his life. When we know our final goal, we can at least ascertain with certainty the direction in which to proceed. Furthermore, if the final goal of human life relates to values that are characteristic of all philosophical knowledge, it will then afford us the critical criteria for making the right choices in regard to action, by classifying them as bad, good or better, a rank-ordering in terms of preferences. Since the freedom to choose which action to take at any given moment is a direct consequence of human self-awareness, reference to a set of established criteria will be useful in proceeding towards the goal without needless faltering. The mission of philosophy, therefore, asks for the formulation of an end or ends before we can address the moral evil.

We have so far spoken of the end towards which man should strive without specifying what, in fact, it is. Obviously, this ideal of life did not dawn on man in one fell swoop but crystallized over a period of time. Just as in the scientific endeavor, in which man has proceeded with a preliminary understanding of a new observation without waiting for a formal theory to account for it in an elegant and over-arching fashion, so also in the investigation of the goal or goals of human life, man has taken bold steps forward without waiting for a complete understanding. It is important to reiterate that it is man's self-awareness that provided the impetus for the philosophical journey, not self-preservation, which was triggered by physical evil, which itself has been greatly reduced by scientific and technological progress.

Man's self-awareness very soon convinced him that others also possess a similar faculty and therefore they are also endowed with the ability to make choices in their actions pursuant to their likes and dislikes. This fundamental understanding, though based on inference, led him to formulate some guiding principles for the conduct of his life, which would ensure that altruistic actions should take precedence over self-serving ones. However, this lofty goal is more easily stated in theory than achieved in practice: man's natural propensity is to ascribe more importance to his own preferences than to those of others. In other words, there is always an inherent tension between the concept of self-love and universal goodness. If the balance should shift towards the latter, the moral agent is prepared to sacrifice himself. How to solve this riddle comes within the domain of ethics which is a branch of philosophy, a subject which we have discussed elsewhere.

The bottom-up approach for the philosophical problem, where the analysis proceeded from an examination of the human condition, can only proceed thus far. The conflict between self-love and universal goodness cannot be resolved in the absence of a final goal for spiritual evolution. There is no unique formulation to this philosophical problem, and that is why there are so many philosophical treatises in existence. In this section, we have formulated the goal in the light of vedic philosophy which we have discussed earlier. Suffice it to say that its main concern is to probe deep into the true nature of self that we have encountered in the concept of self-love. In our common-sense observation, we find that the word self does not have a fixed meaning and can acquire different meanings depending on the context in which it is used. Awareness of the self also means a simultaneous awareness of the not-self which is complementary to it. The not-self could be the human body when one says that he wants put an end to it; in this instance, the self is associated with the body. In another example, the not-self can stand for the mind when one says that ' I am thinking', in which case the self is associated with the mind. In the case of self-love and universal goodness, the self is associated with other selves.

These examples illustrate the relative meaning of self. Vedic philosophy postulates an absolute self which is not at all associated with any features.

Who am I is the basic question that Vedic philosophy addresses. The final result of the search is that the true and constant self is the highest mode of consciousness in man. This transcendental consciousness is denoted by various terms: *ātman*, pure consciousness, Self with a capital S, eternal witness, Divine ground etc. Vedic philosophy also postulates a dual concept called *Brahman*, which is the ultimate truth behind the external universe, also known as the cosmic consciousness. The master stroke of this philosophical development is that an identity between *ātman* and *Brahman* is established on the basis of their duality. The tug -of -war between self-love and universal goodness is resolved on the platform of the pure self, the ātman. This ideal is at the highest platform compared to what the relative self in self-love can ever reach. Since Vedic philosophy also endeavors to realize the values implied in its formulation, it signals the direction which moral training should take in order to bridge the gap between compassionate thought and the actions which are not always in conformity with it. Even if we concede that the ideal of the highest transcendental sate cannot be reached by the majority of people, the practical benefit is that it provides a set of regulative principles for conduct. That is why Vedic philosophy places a premium on methods for strengthening the intellectual understanding of the highest ideal through prayer and meditation, and concurrently, emphasizes the importance of cleansing the doors of perception by leading a moral life. Furthermore, it points out that contemplating the ultimate truth and pursuing moral life reinforce each other; one without the other would be incomplete.

Our purpose was to show how the study of philosophy can motivate people to lead purposeful lives. We have analyzed some of the consequences of self-awareness, which is an unique endowment of man. We have taken the discussion to the point where the formulation of a philosophical ideal becomes an inescapable necessity. Our choice of the ideal based on Vedic philosophy is consistent with the main theme of the book. However, we conclude the discussion without going into a more detailed discussion of the ideal since it is outside the scope of this section.

3.2 Respect for other religions

Since the purpose of religion is to show the path for a seeker to attain inner peace and enduring happiness by instilling an unshakeable faith in the existence of a supreme being guiding the affairs of man and nature, one would naturally expect that it would , at the very minimum, reject restrictive and hostile emotions of every kind. However, in practice, one finds that religion has not met even its minimum expectations as is evidenced by the conflicts that beset major civilizations of the world. What is more alarming is that the locus of hatred encompasses religion also, the very institution that is meant to mitigate it. There are examples galore of militancy in the practice of religion, so much so that clashes between civilizations are deemed as a distinct possibility as men pursue political power. Some religions are perceived to be more militant than others because of the prominence they give to fundamentalism emanating from narrow interpretations of their scriptural texts. On the one hand, we take pride in the fact that the tremendous advances in transportation and communication have transformed the whole world into a global village; on the other hand, we are also cognizant of the mental distance that separates the practitioner of one faith from another. Electronic proximity achieved through the internet or the apparent shrinking of time zones due to the speed of jet travel does not necessarily ensure proximity on the mental and physical planes. In addition to the problems posed by the lack of tolerance between religions, we also encounter a different kind of hatred on the part of atheists when they view religion with open disdain.

The question of mutual acceptance of religions has been debated throughout the ages in order to fathom why non-acceptance creeps into religion at all. The need for inter-faith dialogue in order to reach some amicable understanding has gained momentum with the advent of the new millennium. We find that every religion preaches universal love towards all mankind and makes it incumbent on its devotees to practice good ethical behavior as a matter of duty. But, it is well known that there is a wide gap between precepts and practice when it comes to the question of cultivating genuine regard for other religions. Only a critical analysis can reveal whether there is anything in the teaching of a religion itself that gives room for intolerance, even indirectly. In order to be on a solid footing, respect for other religions should not come merely as a matter of good protocol, but it should arise out of a deeper conviction of one's own conceptual framework regarding the totality of the truth that is taught in one's own religion. We shall now examine some of the ways in which we can reconcile the truths propounded by the various religions.

One way to avoid being fanatical about one's own religion is to realize that the truth taught by it cannot be asserted with complete certainty on account of its transcendental character. Transcendental truth cannot be verified by the methods available for verification of empirical truth because the former is beyond the pale of human experience. This conviction should instill a degree of healthy skepticism towards one's own understanding of religious truth. This, in turn, should also inject a modicum of respect for the teachings of other religions, which also have identical limitations. Truly speaking, however, this kind of skepticism, arising from a lack of certainty about the doctrine to which one adheres does not pass for religious toleration. If one starts from the premise that there is uncertainty attendant on the truths proclaimed by all religions, what one is really suggesting is that the object of intolerance is not at all present, and so the question of religious intolerance does not really exist. Furthermore, the premise conveys the absurd idea that those who believe about their religion's truths are, by definition, intolerant.

A second way of encouraging religious toleration is to grant that there could be certainty about religious truths but they can at best be apprehended by only a few highly evolved individuals, the sages and mystics of religion, and so they need not concern the bulk of the humanity, which has neither the ability nor the desire to reach such heights of spiritual vision. But this type of toleration, while it concedes that it is humanly possible to attest to the certainty of religious truths, it nevertheless jeopardizes the role of religion itself by denying for the spiritual advancement of every human being, which is a highly cherished egalitarian principle. Accordingly, we have to discard this second argument as well as being legitimate grounds for inter-religious understanding.

We shall now present the prevalent viewpoint within Hinduism which serves to promote not only religious toleration but also mutual acceptance. It holds that all religions are equally valid, and it is utterly unreasonable to doubt the authority of any of them or to prefer one over the other. The revealed knowledge of all religions is divinely inspired, and all of them are known to have given spiritual solace to their adherents. Consequently, whatever the the seeker chooses, he is sure to be following the correct path. One can realize the ultimate truth by having recourse to one of several approaches. This is the bond that links all human beings. The differences that exist between religions are attributed to differences in evolution on account of circumstances such as age and country, and race and temperament. They are confined to the externals of a religion, namely, the rituals and modes of worship; in particular, they do not suggest discrepancies in their inner truths . The external variations are conspicuously present within Hinduism itself; the manner of worship, say in a Ramakrishna Mission, which places more emphasis on meditative practices, is very different from the temples where the dominant mode of worship is through rituals and Vedic chanting. An outsider to the religion might even wonder whether it is the same religion that the two groups are practicing. Because of such wide differences in the practice of their own religion, Hindus are not dismayed by the unfamiliar practices of other religions however different they are from their own.

The seed of intolerance is sown when a religion asserts that its claims are true to the exclusion of the rest. It is this ill-informed conviction about exclusivity that begets bigotry, and taken to its extreme, it results in religious fundamentalism and sometimes militancy. Religions based on spiritual revelation to a prophet, combined with a literal interpretation of their scriptures, are particularly vulnerable to the claims of *exclusivity* and *uniqueness* making it difficult for them to admit that there could be many paths to the divine. If a clash of civilizations, something that the doomsday political scientists envisage, is be avoided, it is necessary to eliminate the feeling of exclusivity that is present in some of the world religions. Hindus are convinced of the non-exclusivity of religions and thus have no need for proselytizing. Religious tolerance has characterized the mainstream of Hinduism throughout its history. It is this dominant force that has been able to mitigate the intolerant fringe movements within Hinduism that have surfaced in recent years. A conviction should grow that the tolerance with which religious truth is pursued is as important as the pursuit of the truth itself. The following quote from Hiriyanna [21] highlights the universality of religion: no divisions are so sharp as those caused by religion and it is equally true that no unity is so strong as that following a recognition of identity in religious aim which is the supreme aim of life.

3.3 Śravaṇa, Manana and Nididhyāsana

We said earlier that it is *avidyā* or spiritual ignorance that is responsible for diverting our attention away from the true spiritual reality that resides within us. The goal is therefore to devise ways and means to overcome this barrier.

Mere intellectual understanding of the problem is not enough for overcoming spiritual ignorance. For analytical purposes, the procedure is described in three discrete steps although, in reality, they overlap each other. In the first phase, various verbal models are devised to delineate the different aspects of the one supreme reality and extoll its intrinsic qualities. The second phase consists of providing an intellectual understanding of the abstract principles of *Brahman* and *Ātman*. The third and last phase deals with the practical aspects at an experiential level of total fulfillment. This asks for a proper assimilation of the ideas of the previous two phases.

Śravaṇa refers to the study of the scriptures that is concerned with the aspect of hearing about the Truth from qualified spiritual guides. This is the first phase. The Truth is about the two principles of *Ātman* and *Brahman*, and according to the nondualistic interpretation of the scriptures, they are identical. In theory, it is preferable to hear about such lofty matters from a guide who is an embodiment of spiritual perfection, but that, unfortunately, is a requirement that cannot be easily satisfied. The practical alternative at our disposal is to identify a teacher who is well advanced in spiritual practice and about whose teaching we can associate a degree of certitude. Even this is not an easy task and can involve considerable searching. It is not only necessary that the teacher should be qualified but also that he should be the one with whom we have a chance of establishing instant contact. It is a common experience in secular studies that we find that even an outstanding professor is not rated equally by every student. With some, we instantly click, and with some others we don't. Either scenario is not necessarily attributable to prejudice or other negative feelings. This response is greater with spiritual guides where the communication takes place at a much deeper level of one's being. Since the theme of spiritual instruction is quite profound, it will generally need constant repetition before the message leaves a permanent imprint on the mind of the uninitiated.

After *śravaṇa* comes *manana* which is the stage for intellectual understanding of the truth that we have repeatedly heard from various angles. Since Vedas constitute the *pramāṇa*, the testimony of the existence of the supreme reality, the line of reasoning is expected to follow within the confines of this testimony. Reasoning without this guideline is not recommended since it may lead one to erroneous conclusions because of the possibility of indulging in arbitrary speculations. It is the stage where one acquires the indirect knowledge about the Truth, beyond a shadow of doubt. It is indirect because it still remains a mental construct unverified by experience, and only experience can serve as the final arbiter in settling matters which relate to the spiritual realm.

Nididhyāsana is the final stage of direct contemplation about the Truth itself. This is the stage of spiritual practice which includes deep meditation. The chief characteristic of this stage is that it is marked by a steady progression at the experiential level. The spiritual aspirant would by then have progressed in reducing the opacity of his ignorance and gained purity of mind (*cittaṣuddhi*) and one-pointedness (*ekāgratā*) in the pursuit of his goal. Constant inquiry about the Self, meditation, and skill in action in worldly matters are all prerequisites for this third and final stage of spiritual practice. Since most of the time we are engaged in worldly activities , it behooves one to act in such matters in a way that is conducive to accelerating towards the spiritual goal. This reasoning implies that the attitude that we bring to action is very important in deciding its merit. The sacred text of Gītā is a treasure house of information about the practical techniques that one can cultivate for getting more and more adept in the exercise of skill in action.

It should not be construed that the three stages of *śravaṇa*, *manana* and *nididhyāsana* should come one after another in an orderly manner. Although they are discussed separately for analytical purposes, they do overlap with each other in a devotee's life. Practice reinforces theory which, in turn, will further reinforce practice. In fact, even an iota of real experience which gives one a fleeting glimpse about the Reality has the power to dissolve all doubts about its existence. The spiritual aspirant is more likely to concentrate more on practice from then on rather than indulging in endless discussion about the theoretical aspects.

3.4 The Karma Doctrine

The *Karma* doctrine has influenced the lives of the Hindus from time immemorial, and even today its impact on the society is all-pervasive. In fact, it is difficult to fathom the Hindu mind without a proper understanding of the *Karma* doctrine and its many ramifications. One could go even further and state that the doctrine is common to all religions indigenous to India such as Jainism and Buddhism. Only, the philosophical school of Cārvāka, which is based on materialism wherein only the reality based on the data provided by the senses is recognized, does not subscribe to the *Karma* doctrine.

The *Karma* doctrine has two principal features to it. First, it believes that the law of causation and strict determinism that are encountered in many aspects of the physical world are also applicable to the realm of human conduct. Second, it emphasizes that there is perfect moral retribution underlying the law of causation. Without a clear understanding of the implications of the second feature, it is possible to come to wrong conclusions about the doctrine itself.

The inference from the two features of the doctrine is that the moral realm is completely preordained. It does not follow, however, that it amounts to a doctrine of fatalism with all its grim pessimistic implications. There is absolutely no suggestion in the doctrine to entertain a helpless attitude of total resignation to what will inevitably happen in the future, which is what fatalism would imply. On the contrary, an individual's thoughts and actions are held to be the sole determinants in deciding the course of his spiritual progress, which is a philosophy that ensures a robust optimism for life. The doctrine places a premium on the value of human freedom whereby the individual is endowed with the conscious choice to either accelerate or decelerate the pace of his own spiritual evolution to realize the divine element in him. The importance of this most valuable asset that a human being possesses is highlighted by contrasting it with the lack of freedom that is characteristic of the animal species which has to totally depend on instincts and so, on the guidance of Nature for their spiritual evolution.

The emphasis on freedom at the individual level to think and act might seem to contradict God's qualities of omnipotence and omniscience. This is the problem of free will which we have discussed earlier. The Hindu viewpoint does not see any inherent contradiction in the situation since the worldly reality of man is entirely different from the transcendental reality of God. Since they refer to two different planes of existence, it is believed that there could be no confusion introduced because of the problem of free will. The nondualistic interpretation of the *Upaniṣadic* message simply dismisses the difficulty since it recognizes only one ultimate reality from whose vantage point the worldly reality is only an illusion.

Despite the promise of moral retribution, which is the second feature of the *Karma* doctrine, one witnesses a great deal of sorrow and suffering, pain and tragedy, most of which happens for no discernible reasons. We witness all the time the grotesque inequities of life for which we find no rational explanation to console ourselves. Faced with such direct evidence, it becomes difficult to be convinced of the justice of the doctrine without any further explanation beyond what is implied by its two features. This philosophical paradox is resolved by invoking yet another concept, called *saṁsāra* which is also an integral part of the *Karma* doctrine.

Saṁsāra is the continued existence of a soul (*jīva*) in a succession of both past and future lives. The theory of transmigration is absolutely essential for validating the *Karma* doctrine, and the evidence for rebirth (*punarjanma*) is entirely based on the declaration of the scriptures. As a consequence of the concept of *saṁsāra*, it is possible to offer an explanation for the grave injustices of life without invoking an external fate to account for them. Any such recognition of an outside influence would have acted as a constraint to man's freedom that would be contrary to the fundamental tenets of the *Karma* doctrine.

The concept of *saṁsāra* entails the idea that the reason why freedom was exercised in the way it was done in the present life for a particular act can be traced back to an event in one of the past lives. How far back do we go for seeking such an explanation? This question remains unanswered because of our ignorance of our past lives, and consequently, we have no means of identifying the time in one's past life that would be responsible for generating a cause for a consequence which has sprouted in the present life. All that information is believed to be meticulously recorded in an invisible log book. Asking for such information is meaningless since the soul is at no time free of its stored impressions (*saṁskāras*) that are accumulated by the interactions with the external world in a series of lives. Consequently, *karma* is beginning-less, a concept called *anādi.*

The *Karma* doctrine appears so autonomous that it does not even acknowledge the role of God. The self-contained explanation is that the cumulative impressions which are etched in an individual soul over a series of lives lived in the past influence the course of the present life, and that, in turn, will affect the future in an endless cycle. But the system of rewards at any stage is based strictly on the concept of moral retribution which is what gives the powerful encouragement for moral growth. The ultimate reward, of course, is to break the chain of *saṁsāra.* That is possible only when the *Karma* doctrine becomes inoperative which happens when there is no more rebirth for that individual soul. This, in fact, is another explanation of spiritual realization whose attainment very much depends upon the tact with which one exercises one's human freedom which is assured by the *Karma* doctrine.

The *Karma* doctrine can be explained in an alternate way using the terminology of the physical sciences. If attention is only limited to a single span of life, we do not witness the forces of determinism and moral retribution at play. The affairs of life appear to be utterly chaotic for the majority of mankind. Accordingly, the events of a single life can for all intents and purposes be viewed as a random phenomenon where there is no strict cause and effect relationship. On the other hand, it is the concept of reincarnation that enables one to offer the philosophical explanation that the events of life are indeed deterministic over the time-scale of spiritual evolution of the individual soul. We are only deluded as to the real nature of the phenomenon in a single life span because of our ignorance of what has happened to the individual soul over the endless cycle of rebirths up to the present life. In other words, the doctrine appears random only because of our inability to discern the deterministic chain of events preceding one's life. Recalling the famous statement of Einstein with regard to the true nature of quantum mechanics, we can say that whether God plays dice with nature or not, the *Karma* doctrine informs us that He certainly does not play dice in the case of the spiritual evolution of the individual soul. The above line of explanation is only meant to provide yet another insight into the nature of the doctrine. We say this guardedly because we cannot extrapolate a scientific argument into the philosophical realm and claim its validity even there.

There have been attempts to probe into the question of past lives on a systematic basis to see whether we can gather concrete scientific evidence for the belief. The stories of individuals who have recounted their experiences of their past lives have provided the impetus for this line of research. Specifically, the motivation for such research is not at all connected with the *Karma* doctrine *per se.* Freeman Dyson has speculated on the rich possibilities of future technologies concerning life and mind. It appears that in the future there will be the scientific curiosity to develop powerful technologies to delve deep into the inner recesses of the mind.

Already we have a good scientific understanding of life on the basis of molecular biology. The possibility of correctly mapping genetic blueprints, the DNA sequences, is at hand. When that knowledge becomes available, one can even expect clones of the genetic blueprints of deceased persons whose records are available. This technology would allow us to essentially resurrect the ghosts of our ancestors. The next big step would be to try and read the memory traces, the *saṁskāras* of individuals. Right now, it seems like a science fiction even to think of a technology to read and write the memory traces , but the day may come when that will happen. After all, says Dyson, we take the ability to recall the events of our childhood without wondering about the mystery surrounding the function of remembering. The technology needs to externalize this ability, for which we need to know where and how to capture the information. It is only with the mastery of such technologies that we can begin to probe the question of past lives in a scientific way. For now, the *Karma* doctrine has to be accepted on the basis of philosophical reasoning and the verbal testimony of the Vedas.

3.5 Dharma, Artha, Kāma, Mokṣa

Dharma stands for Ethics, *artha* for matters that provide security in life, *kāma* for sensual pleasures, and *mokṣa* for liberation. These four factors are designated as *puruṣārthas* which means ‘ aims in life’. These four goals can be further classified into two sets: *artha* and *kāma* in the first set, and *dharma* and *mokṣa* in the second set. The reason for this classification is that the goals of the first set are not only applicable to human beings but also to all of the living species on this planet. The second set is exclusively meant for man because of his possession of the unique faculty of self-awareness.

Even in the context of the first set, however, there is an important difference between man and the rest of the living species. The non-human species pursue their goals for security and pleasure through instinct only, whereas in the case of humans, the activity is subjected to the power of discretion. Depending upon the interpretation of this higher unseen force, instinct can be seen as something that is pre-programmed either by God or through the governance of a cosmic law. For pursuits guided by instinct, there are natural constraints. For instance, an animal does not get greedy and claim more than its due share once its real needs are satisfied. Since humans, on the other hand, are vested with the power of discretion, the sky could be the limit for the pursuit of their security and pleasures. The rise of the ' economic man' and the reckless manner in which human beings are exploiting the resources of the world should make this point clear.

Humans need some self-imposed constraints so that they act in such a way as to satisfy their legitimate needs of security and pleasure without, at the same time, getting into conflict with the needs of the world at large. The discussion of such constraints will lead us to the domain of ethics. Ultimately, the recognition of the need for constraints means imbibing a value system that would enable humans to exercise control over both means and ends in such a way that they satisfy their legitimate needs of security and pleasure and at the same time work in consonance with the needs of the rest on this planet. This value system is called ethics, and it can be considered in two parts. First, is the ordinary ethical values which are based on the principle of reciprocity; what this means is that you do unto others what you expect of them in return. We don't inflict physical or mental injury on others since we do not want them to behave in such manner towards us. This set of ethical principles, which are upheld by everyone irrespective of whether they subscribe to religious beliefs or not, gives rise to a list of do's and dont's.

There are, however, complex situations where satisfaction of the first set of ethical principles will not automatically resolve the problem of ethical uncertainty. In the name of honesty, we cannot, for instance, naively pass on sensitive information to some one whom we know is out to cause incalculable damage. The story of Arjuna on the battlefield just before the commencement of the Mahabharata war is a classic example of an ethical dilemma where the outcome of action was shrouded in moral uncertainty. Such situations ask for reliance on a second set of ethical principles, called *dharma*, which is based on the scriptures or on religion. Broadly speaking, observance of *dharma* ensures that the individual action is always in consonance with the cosmic law operating in the universe. We have stated earlier that the need for ethics does not arise in the animal species since what is meant by action according to instinct is that it is automatically in consonance with the cosmic law. In the case of humans, however, because of the ability to act one way or the other guided by discretion, there is always a possibility of losing synchronism with the cosmic law. The concept of *dharma* ensures that there is guidance to act properly in difficult situations where there is a possibility of intangible outcomes. It also provides the constraints which are absolutely necessary for the exercise of the earlier two aims of *artha* and *kāma*. The twin concepts of *puṅya* (religious merit) and *pāpa* (religious demerit) also emphasize that the results need not refer to the present life only.

Our earlier discussion of the *Karma* doctrine should throw further light on these concepts.

The concept of liberation (*mokṣa*), which is the fourth in the list of aims for life, appears in various forms in all the metaphysical doctrines that are discussed in the later chapters, and so we shall not discuss it here except to say that it is also exclusively a human pursuit.

3.6 Ethics

The subject of ethics is customarily discussed in conjunction with metaphysics. While this approach lends itself to a comprehensive analysis, it suffers from the usual drawbacks of all top-down treatises, in which one has to traverse a long distance from the premise before arriving at the specific problem of interest. In the case of ethics, the premise of the metaphysical problem is a realm of abstractions dealing with the whole of existence: a subject that would not be of interest to most people. It would place an unnecessary burden on the student to suggest that knowledge of metaphysics is a prerequisite to understanding the main principles of ethics. The difficulty underlying the methodology of the top-down approach is compounded by the prevalence of numerous metaphysical theories; each metaphysical theory essentially gives rise to its own special version of ethics. Such being the case, it is impossible to expect universal agreement on any one of them because of the lack of specific guidelines for selection. In practice, however, the diverse solutions do fall within a narrow band, thus giving further credence to the possibility of approaching the problem of ethics in a more direct manner than what is rendered possible through a top-down methodology.

Ethical considerations arise when one has to select a right and good course of action from amongst several alternatives. A careful and systematic analysis of the basis for the contradictions underlying the alternatives assists in the formulation of concrete methods for an unambiguous choice, and constitutes the very essence of the bottom-up approach. This empirical approach, which is the focus of this section, has the possibility of gaining wider acceptance since it does not necessitate an initial commitment to any single metaphysical theory. The need to understand some basic metaphysical concepts for forging a link between ethics and metaphysics, however, shows up only in the final stages of this analysis and need not be initially anticipated by the reader.

The need for good moral conduct arises in man because of his special faculty of self-consciousness. We shall not digress into a discussion of the subject of consciousness, which is of current scholarly interest in several modern disciplines including artificial intelligence, which is a sub-discipline of computer sciences. For our purpose, it will suffice to take the commonsense understanding of the word consciousness that is corroborated at an experiential level. Self-consciousness, or self-awareness, is a unique feature of man, which distinguishes him from the animal kingdom. Socio-biologists inform us that although one can detect a glimmer of self-consciousness in some highly evolved animals, the difference in degree, however, is so extraordinary that the unparalleled ability of man in this regard is, for all intents and purposes, a difference in kind rather than of degree. The spiritual literature on this topic leaves no room even for the least amount of ambiguity as evidenced by the declaration that self-consciousness is the unique endowment of man. Ethical conduct is, by definition, the actions that can follow from the well-trained and highly-disciplined prompting of self-consciousness in order to achieve the good of the individual concerned as well as of the people with whom he is interacting.

The very first observation about self-consciousness is that it causes man to feel quite apart from the rest of the universe. In scientific parlance, this is called the observer and observed relationship, which is fundamental to the success of most of the scientific enterprise. Self-awareness gives rise to an experience of duality between oneself and the rest of the universe. Before we proceed with the discussion of ethics, it is necessary to have some preliminary idea of the sense in which we use the word self in the hyphenated word self-consciousness. There is a wide range of meanings between the two extremes of the gross self and the subtle self. Before reaching the subtle level, the self can be associated with any of the features of the body, mind, and intellect; the word I can be used in conjunction with any one of them. In most human experience, the self is always connected to some feature or the other. At the other extreme, when the self is completely stripped of all its features, which is a metaphysical concept, the subtle level reveals itself and it is variously described in the vedic literature as Self with a capital S, pure consciousness, absolute consciousness, divine consciousness, etc. We are not immediately interested in this subtlest self, and our discussion will focus on the self of our day-to-day experience, which is associated with some feature of the body or mind.

For the discussion of ethics, it is enough if we can conceive of an I that can take into account the idea of morality, that is, of one's ability to choose and act freely. That ability is possible if the mind is vested with a sense of discretion, unlike a robot which is programmed and thus driven according to a prior plan. The human mind is able to consider alternate courses of action in order to gratify one's likes and avoid one's dislikes. Satisfaction of one's likes results in pleasure, and actions based on dislikes are avoided in order to prevent inflicting pain on one-self. At this stage, we take note of another important feature of self-consciousness. An individual can, through analogical reasoning, arrive at the conclusion that his fellow human beings also possess self-consciousness, and, consequently, infer that they would also like to act in accordance with their own special set of likes and dislikes. Furthermore, this analogical inference is universally applicable. Thus, the moral attitude emanates from an extremely egalitarian principle that is intrinsic to the mind.

Self-consciousness, with its attendant feeling of duality and equal concern for all human beings, engenders the inescapable need for a principle of reciprocity in man's social behavior. Person A is expected to behave towards B in the same manner as A expects B to behave towards him. In fact, this is one of the basic tenets of criminal jurisprudence. Since we agree that the desire to seek pleasure and avoid pain is common to everyone, ideally speaking, we should do to others what gives them maximum pleasure just as we expect them to harbor similar feelings towards us, backed up by appropriate actions. The other side of the equation is that we should not inflict harm on others because that would deny them their right to exist without pain. We will now concentrate only on the positive aspect of the principle of reciprocity that deals with actions leading to maximizing pleasure. This is also the underlying argument of psychological hedonism.

The principle of reciprocity in our relation to our fellow man bestows equal importance to both parties with the compelling necessity of simultaneity. Unfortunately, our distinct idea about the desirability to do good to our fellow man, dictated by the global consideration of reciprocity, does not suppress our urge to assign a higher priority to our own personal interests, thus giving rise to a serious internal contradiction. We do not always wish to obey the prompting of our self-consciousness; the mind has a double nature: one that recognizes the validity of equal concern for everyone, and a second that resists the first by seeking the satisfaction of our own interests. We sadly come to the conclusion that the idea of the self is not so simple as we thought at the beginning of our discussion. We can now talk of a higher self and a lower self , which are opposed to each other; the former reminds us of our obligation to others and the latter urges us to take care of our interests first and foremost even at the cost of skewing the equation of reciprocity. This is the second implication of self-consciousness, next only to the feeling of duality where the feeling of dichotomy of the self was not envisaged. The contending forces between the two selves give rise to a moral constraint which has to be overcome in order to restore the spirit underlying the principle of reciprocity.

Overcoming moral constraint means restoring the primacy of the higher self, and this is where the concept of duty as commonly understood arises. Three of its principal features can be identified: a) in our interactions with our fellow human beings, that is, in the social realm, actions are directed towards others; b) what is required is conscious action dictated by our self-awareness, and not by instincts as in the case of animals, which also work for the welfare of others guided by their herd-instinct; and c) we assign little or no tangible rewards to ourselves in the performance of our duties, which means that our natural inclinations may not necessarily be satisfied. What is prescribed for the removal of the moral constraint in terms of duty might appear as unduly harsh, but this is what it takes to turn a personal inclination into a personal obligation, which is an essential step in ethical progress.

Duty, if it is merely directed solely to the welfare of others, would not contain a real incentive for action despite the deep-seated conviction in the principle of reciprocity which guarantees that similar duties when carried out by others towards the welfare of the agent would satisfy his desires and inclinations. To perform his duty with conviction, the individual must be able to reap rewards from his actions although the rewards may not satisfy his immediate desires. Thus, duty has a two-fold purpose: first, an external aspect directed towards a fellow being, and secondly, an internal aspect directed towards one's own benefit. The internal aspect consists in the gradual development of one's own character as a result of performing actions in the right spirit. For a duty to qualify as a moral action, it should satisfy both the conditions. The simple act of helping someone in distress will not qualify for moral action if it is not done in the proper spirit. One of the classic examples is that if a person saves a drowning man merely because he is his debtor, the outward form of valor does not constitute a moral action because it is void of the inner spirit. One is also reminded of the example given by Samuel Johnson in which a rich man throws a coin towards a poor man in order to hurt him; the poor man is grateful for receiving the coin, but the donor did not give it in the right spirit, so the action does not qualify as a moral action.

Our discussion so far has implicitly assumed that the concept of self whether it be of the agent or of others pertains to the instant of time at which the duty is being performed. For a fuller discussion, it is necessary to remove this restriction and consider the self as it evolves from birth to death. In other words, it is necessary to bring in the factor of time, over which a metamorphosis of the self takes place. The bundle of likes and dislikes which are at the basis of the principle of reciprocity will not remain constant throughout the life-time of any individual; thus, there is a need for a reinterpretation of the principle when an interval of time is taken into account.

In passing, we take note of the fact that we have set aside consideration of the self in its subtlest state, in which attraction and repulsion due to likes and dislikes are totally absent, because this is associated not with time but with eternity. The word eternity, as it is used here, does not mean from everlasting to everlasting, but that which transcends time. This is where there could be differences in ethical systems associated with the particular metaphysical theories. Furthermore, there is no simple way of resolving the differences since the transcendental realm cannot be reached even when we take rationality to its very limits. Fortunately, we do not have to delve deeper into the subject because our present thesis deals with empirical ethics, in which time is the essential factor, and because the self of our discussion is always associated with some external features.

The psychological experience of time passage is fundamental to every human being. We have the concepts of the past, present, and future, and because of the putative unidirectional flow of time, the past merges into the present and the present into the future. The notion of becoming is paramount here. Let us focus on the impact of time on self, not of the agent, but of the person he is interacting with. If, according to the principle of reciprocity, the agent has to indulge in such activities that would maximize the pleasures of the second person, the uncertainty arises as to which moment of his life needs to be considered since his likes and dislikes do not remain constant throughout his life time. Consequently, one would per force rather look for an enduring pleasure than an ephemeral one. Such an unalloyed pleasure is called happiness. The principle of reciprocity is now restated in terms of the maximization of another's happiness instead of pleasures. The next question that arises is: how do we enrich the meaning of the principle by taking the whole of humanity into account? Theoretically, it would mean that man should strive for the greatest happiness of the greatest number, i.e., utilitarianism.

We can now bring in the impact of the time interval on the self of the moral agent, a consideration that we had set aside in our discussion leading to the concept of utilitarianism. The principle of reciprocity would be devoid of its true meaning if a real incentive for action were denied to the agent. He should also realize that his exercise of duty to others should be such that it has the potential to increase his level of happiness in an enduring way. Furthermore, he should be conscious of the fact that while the happiness of others is only an intellectual conviction as far as he is concerned, it is, on the other hand, an emotional experience when it comes to his own case. No one can truly experience some one else's emotions. All said and done, each person has to work for his own ethical improvement; others can at best assist one in the process. Consequently, there is a difference in the type of happiness that results through the application of the reciprocity principle in its broader setting.

The happiness of the agent is placed on a higher footing altogether than that of the person he is interacting with; the happiness of the agent can be considered as a higher good than that of others. It results in the building of one's character; the litmus test is when that which was once considered as a moral constraint is gradually felt as a moral obligation.

The discussion on the higher good needs some more elaboration in order to make its meaning clearer. The goal of achieving perfection of character is equally attainable for everybody. While this ideal has clarity when applied to one's own self, it is obscure when it comes to the perception of a similar ideal present in others. All that we can say is that moral action should not only result in the achievement of a spiritual goal in the agent but also indirectly assist others with whom we interact in the realization of their goals and directly assist them in the realization of the lower good, i.e., their specific pleasures. This should be possible since one cannot conceive of an internal contradiction in the two pursuits: A can work for his own higher good, but he can only be certain of achieving the lower good for B as a result of his moral action and, in addition, indirectly work towards the achievement of the higher good of B.

As we have stated earlier, for an action to qualify as a moral one, it is not only its outward form that is important but also the inner spirit with which it is done. When so much emphasis is laid on the inner spirit of a moral action, it is necessary to examine some objective criteria for establishing the worthiness of internal standards as they would otherwise be deemed purely subjective. The time-honored way of providing correctives for maintaining internal standards is making use of the yardstick of social approval. Social approval hinges on the accordance of long traditions and current practice and, in addition, is approved by the reflective minds of the society.

Under normal circumstances and in a majority of cases, both the internal standard of the individual and the moral judgment of the society coincide, but they can vary when a society undergoes rapid changes. It is well known from experience that societies change their values in the course of history. To cite only one example, a generation ago, one could not have openly spoken about gay rights because of the taboo that was placed on such sexual behavior, but it is now a matter of public discourse because of our understanding that it is a sexual orientation that has existed from time immemorial. Examples such as this abound in every society, and so it is necessary to examine what causes ethical dilemmas for an individual. Using the terminology that we have developed for our exposition of empirical ethics, we can say the instability in social standards is caused by changes in our conception of the higher self. It can take on the meaning of a tribal self of a primitive society or, at the other extreme, the subtlest meaning afforded in philosophical and religious discourses. The double nature of the mind results from a fixed self (lower self) responsible for seeking pleasure and avoiding pain, and a dynamic self (higher) depending on the evolution of the society as a whole. But, whatever the stage of evolution of a society, all other implications of the principle of reciprocity are still valid. The ethical problem, in the ultimate analysis, remains an individual one because of the internal contradictions posed by the double nature of the mind, albeit with the changes that occur with changes in social standards.

Is there a fixed absolute value for the higher self ? The answer takes us to an understanding of metaphysics. If we are merely interested in empirical ethics, which for all intents and purposes should be quite adequate to lead purposeful lives, we need not answer the question. But the incessant search for finding the real meaning of ethics arises from man's insatiable urge to attaining perfection, and so we make some concluding comments on the metaphysical question. We had commenced with the idea of self-consciousness in developing the central concepts of ethics and claimed that there is a self which is stripped of all features. One of the names assigned to it was pure consciousness. This highest self of the double nature of the mind was also referred to as Self. When that level of consciousness is realized, the lower self merges into the higher self and no duality is perceived. This is the essential argument of an Indian metaphysical theory based on nonduality. But the same central truth can be restated in different terms depending on one's preferred metaphysical theory. In any case, this is the stage at which the link is provided between ethics and metaphysics.

We have so far proposed a model as it were for explaining the central concepts of empirical ethics. We do not, however, want to leave the impression that, for the principles to be put into practice, one has to painfully follow step by step the ascending order of the present discussion. If that were so, many would conclude that the discussion was meant only for gaining dry intellectual knowledge and not for putting that knowledge into practice because of its hopelessly impossible demands. The real story is more optimistic than that. What is initially required is a concrete step by which we can activate the whole process of moral improvement in a simple way that does not place undue burdens. This is done in a two-stage process: first, by acquiring through prayer and meditation at least a fleeting experience of the highest self that resides in every one, and secondly, by making life-long use of this repetitive experience to sustain the entire process of moral and ethical cleansing in a systematic way. If the process does not trigger higher levels of happiness, one can begin to suspect that there is something wrong with it. The proof comes at an intensely experiential level and hence leaves no room for doubt. The practical discipline is best understood with some initial guidance and cannot be the subject of a mere theoretical discourse, at least not in its entirety.

We wish to point out at this stage that our particular development of the bottom-up approach to ethics is informed by Indian philosophical concepts. Indian philosophy is equivalent to a conception of values because it is not merely interested in distinct knowledge of the ultimate reality, the Self of our discussion, but also in the actual means of proceeding towards this common goal of all mankind in an individual's life-time. This empirical approach to the exposition of Indian philosophy has also provided the conceptual framework for our discussion of ethics. All too often, the New Age gurus who teach Eastern philosophical concepts in the West concentrate only on an intellectual knowledge of the supreme Self followed by meditative disciplines of one type or another. Specifically, they fail to stress the importance of leading an ethical life, which is an integral part of the spiritual journey. Such teachings present such a distorted picture that many westerners have the erroneous impression that practical ethics are not a primary concern of Eastern philosophy. There is a need to correct this misunderstanding because of the wide publicity enjoyed by the gurus.

In summary, we have developed a bottom-up approach for understanding the central principles of ethics. We started out with the observation that self-consciousness is the unique endowment of man and discussed its implication of the duality of the mind. This was followed by a discussion of the reciprocity issuing from man's kinship with his fellows. However, we noted that the obligations implied by reciprocity are not usually met because of the double nature of the self, namely, the lower self and the higher self. It is the preference given to the demands of the lower self based on the likes and dislikes of the individual that disrupts the reciprocity. We have discussed how this moral constraint can be overcome by asserting the role of the higher self in the pursuance of one's duty. We then examined the moral equation when an interval of time was considered. The implications of moral action of the agent when he interacts with another individual were considered in detail. Further, we enlarged the discussion to include ethical considerations for the society as a whole. We also discussed how the social standards change with time and how these could be explained through the model of the higher and lower self. In conclusion, we pointed out how on the basis of an intellectual knowledge of the highest self, coupled with a systematic method for cleansing the doors of perception, one could bring real meaning to the subject of empirical ethics. Finally, we wish to reiterate that no exclusive preference is claimed for a nondualistic metaphysical theory which constitutes the background for the present development. The Self is conceived in different ways by different faiths; hence, there are numerous ways to conceive of its interrelations with the lower self. In fact, such doctrinal differences exist even within a single faith. But what is important for the development of empirical ethics is that it should be finally linked to a self-consistent metaphysical theory. The optimistic conclusion is that, there is a lot of common ground in the conception of ethics of the various faiths.

3.7 Concept of Duty

The background for this section overlaps with the previous section on Ethics. However, this section is included because of the importance attached to the concept of duty in a world where there is so much prominence accorded to individual rights.

One of the fiercely debated subjects in philosophy is whether duty is a means to an end or an end in itself. There are eminent philosophers who take rival positions on this important question, and therefore it is of interest to know the manner in which Vedic philosophy resolves this controversy. For purposes of this discussion, we assume a knowledge of the four goals of human life according to Vedic philosophy: the two secular goals of pleasure and security (*artha* and *kama*), and the remaining two spiritual goals (*dharma* and *moksha*), which we discussed earlier. (See section 3.5). The conception of values as embodied in these four goals provides the framework for investigating the meaning of duty at the very source where the two conflicting views arise.

The relentless pursuit of pleasures, wealth, and success without arriving at a satiation point where one would naturally question the wisdom of pursuing such a life bereft of natural constraints on such pursuits culminates in moral bankruptcy. It is the recognition of the existence of a higher self that makes for the beginning of a spiritual life. However, Moral action, in pursuit of spiritual goals, has a negative impact: it perforce places constraints on the pursuit of gratifying one's natural inclinations in life. Secular goals, it should be emphasized, while essential and legitimate, impede access to the higher goals in life if striven for excessively. What exactly the nature of the limiting constraints should be will become clear as we investigate the positive effect of performing duty.

We shall next examine the type of actions that are deemed appropriate for the pursuit of spiritual goals. It is common knowledge that most actions are performed with definite ends in view, and moral actions are no exception in this regard. No action is possible without the expectation of an end in view; it will always have a definite purpose. It follows from this observation that if moral actions do not result in the achievement of the predetermined moral ends, the constraints placed on the pursuit of secular goals, which constitute the negative effect of moral action, will be utterly meaningless and disheartening since they only serve the purpose of imposing unnecessary hardship on the agent. No rational being will subscribe to a concept of duty that is barren of a positive purpose for himself.

The gratification derived from meeting secular goals without any constraints is called egoistic hedonism. It considers the pursuit of pleasures and security our *raison d'tre.* If such a pattern of life is not considered moral, one could then examine whether egoistic hedonism along with a universal form of hedonism, namely, ensuring the well-being of others exclusively in the secular realm, could serve as the positive purpose of a moral action. Nevertheless, a combined form of these two forms of hedonism is impossible to achieve since one can conclude that one's own interests are always at odds with the interests of others, and as such it is impossible to bring about a reconciliation between the two. Consequently, this solution also has to be rejected in our search for a positive purpose for duty.

We shall next examine whether disinterested altruism could serve as a useful approach. By this we mean that an agent is neither self-serving nor in the pursuit of egoistic hedonism and universal hedonism. Instead, he is totally dedicated to working for the secular welfare of others, denying his own natural inclinations. But this option also has several inherent weaknesses to it. First of all, it is patently absurd to assume that one could unfailingly achieve someone else's happiness at all times through one's actions. What the other man does for himself will also naturally influence his state of mind. Secondly, even if an action is performed with a bad intention towards another person it can, under some circumstances, result in immense benefit to the other; the recipient may be blissfully unaware of the motive of the action of the benefactor and will gladly accept the result if it suits his purpose. In any case, it is safe to assume that the recipient is not at all interested in psychoanalyzing the intentions of the benefactor as long as it suits his purpose. From the foregoing analysis , we come to the inevitable conclusion that even disinterested altruism cannot provide the positive purpose for duty.

Having disposed of the above three possibilities in our search for a positive purpose for a moral action, we finally track it down to the point at which the conflict actually arises. We start our discussion of this case by first recalling our basic observation that self-awareness is a special endowment of man. In our discussion of Ethics, we stated the *principle of reciprocity* in human relations, a principle that logically followed from the very plausible inference that others also have likes and dislikes in the same way that we have ours. We do good to others in order to ensure equally good reciprocal treatment from them. Furthermore, the reciprocity principle based on conscious fellowship remains only in the realm of thought; in practice, it results in a clash between thought and action because of man's natural propensity to favor his own interests. We ascribed this failing to the double nature of the human mind, resulting from the inner conflict between the lower, self-interested self and the higher other-directed self . Thus, the preservation of one's own interests becomes a matter of priority over the higher self which demands one to place others' interests above those of one's own. It is essentially a conflict between the thought that vouches for the soundness of the reciprocity principle and the will that refuses to put it into practice.

The positive purpose of a moral action can be clearly defined in terms of the dilemma posed by the double nature of the mind. It is to eliminate the cleavage that exists between the lower self and the higher self by means of rectification of the will to facilitate the response to the prompting of the higher self. A moral action has two facets to it: first, it should be able to curb the undisciplined impulses that are solely directed towards the satisfaction of the natural inclinations; this is the negative aspect of moral action. The second facet to moral action, which is a positive one, is to ensure that all actions are done with the superior purpose of harnessing the will in the proper direction. It is only in the field of action that one can introduce the constant corrections to the human will and bring it in line with the expectations of the higher self. Although we have discussed the two aspects of moral action separately for purposes of analysis, what happens in practice is that progress achieved in one aspect will also bring about progress in the other because they are mutually coupled. All altruist actions are viewed as opportunities for achieving this higher discipline, the rectification of the will, and with practice, they become second nature to the individual and, consequently, effortless in nature. The importance of doing altruistic deeds is not in the external consequences of the deeds, but in the opportunities they afford for the internal cultivation of the mind. By this observation, we are not suggesting that one should be indifferent to the good of others but are only drawing attention to the primary purpose of heeding the call of the higher self. Constant practice in the rectification and transformation of the will results in a state of happiness that is much superior to the state of happiness that characterizes egoistic hedonism. The latter type of happiness is only transitory in nature, whereas the former is stable and enduring. If this were not true, and if we were only left with the onerous task of satisfying the negative condition, life would be meaningless drudgery. It is the happiness resulting from the positive purpose that provides the necessary incentive for the transformation of one's own personality by performing the right actions.

Normally, when we act, we are conscious of the results we are expecting from it. But, while performing a duty, we cannot always be conscious of the final aim of rectifying the will, which is rather in the abstract realm. The immediate incentive for moral action comes from our ability to distinguish right from wrong and, accordingly, act on what is right. The main reason that we are conscious of the final aim in any single action is that all actions have the same purpose: to strengthen the will so that the gap between thought and impulse in the light of the reciprocity principle is continually reduced, with the resulting happiness that follows from it. But more importantly, the result of a moral action is not external to duty; it is not something that accrues after its performance. The reward for a moral action is contained in the process itself. This is in stark contrast to what we normally mean by an action from which results follow. Egoistic or universal hedonism, which we discussed earlier, have results that are external to the actions performed. Such results may or may not follow the actions, whereas there is no such uncertainty involved in the performance of a duty; it will undoubtedly result in the further rectification of the will. The resulting happiness, being associated with the higher self, is of a much superior and enduring type than the kind of pleasure that one experiences in the satisfaction of purely material goals. In fact, the material and spiritual end result in altogether different orders of happiness.

Reverting back to the controversy of whether duty is a means to an end or an end in itself, we can now suggest a solution to this quandary by pointing out that the conflict is not between duty and end, but between the two orders of end: the lower order, which is associated with the sole satisfaction of the material goals, and the higher order, which is associated with the continual refinement of the human will inspired by the spiritual goals. The school which holds that duty is a means to an end is correct in its assertion when the end is the material. On the contrary, the school which holds that duty is an end in itself is correct when the end is spiritual, i.e., not external to duty. An action becomes moral when it selects the higher order in preference to the lower. By moral life, we mean that one is permanently committed to the performance of duty.

At this final stage of our discussion, we wish to point out where the connection to metaphysics becomes necessary to further explore the complete meaning of duty. We started the discussion on the positive purpose of moral action with the concepts of the principle of reciprocity and the double nature of the mind. The cleavage between the two minds is never completely reconciled unless we have the concept of the highest self, which, admittedly, is a metaphysical concept. It is the

Self of Vedic philosophy that resides in eternity and not in time which represents the highest ideal of life. It represents a state not only of moral rectification but of total moral perfection. Attainment of this ideal is called *moksha*, the highest conception of value for a human being. We did not proceed from this concept of Self because it is possible to deduce the true meaning of duty by confining our discussion to the finite world of time and appealing to facts based on actual experience. Accordingly, the concept of higher self was all that was necessary.

In conclusion, we wish to state that the concept of Self should not be looked upon as a distant and misty ideal. Intellectual knowledge of the Self, combined with a devotion to moral rectitude will enable a person to glimpse the transcendental plane of existence through a yogic discipline. That is the assurance given by people all over the world who have followed this path. It is the momentary experience of Self, the final goal of life, that has the potential to dissolve all manner of doubts about the wisdom of opting for a life dedicated to moral principles. A person who leads such a life will seldom entertain any doubts or be confused about the true meaning of duty; it is both a means to an end as well as an end in itself in light of what we have defined as the two distinct ends governed by physical and spiritual dimensions of man.

3.8 Five Sheaths

Since pure consciousness is a non-manifest reality, even those who have experienced it find it extremely difficult to communicate the exact nature of the experience. The difficulty arises because the experience is incommensurable with worldly realities of the relative realm characterized by spiritual ignorance. Ultimate reality is beyond the reach of intellectual operations, which in turn, depend upon language whose vocabulary is limited only to worldly matters. That is why ultimate reality is ineffable in nature. However, there are various verbal models that can, even with all their imperfections, assist us in orienting our thinking towards that one supreme reality. One such descriptive model deals with the five sheaths that exist between the state of spiritual ignorance and the state of spiritual enlightenment. The discussion of this appears in the *Taittriya Upaniṣad.*

The five sheaths in ascending order from the least to the most subtle levels of existence are *Annamaya*, *Prāṇamaya*, *Manonmaya*, *Vijñānamaya*, and *Ānandamaya*. These are the physical, vital, psychical, rational and blissful sheaths, respectively. The spiritual journey consists of penetrating these successive layers, starting from the physical sheath, until one realizes total enlightenment, which comes after penetrating the final sheath of total bliss. It is called a journey only as a manner of expression since it is not intended to convey the concept of physical distance as we normally understand it. Since the Self is always within us, the destination is not removed from us at all. It is only our spiritual ignorance that puts a distance between us and the destination.

There are further correlations that are established between these five sheaths and the waking, dreaming and sleeping states of consciousness.

Annamaya refers to the physical body, and its main characteristic is the requirement of food in order to survive. This sheath is perceived only in the waking state of consciousness. One's attention to weight or other physical measurements is characteristic of this layer which can be classified as the least subtle level of existence. This is correlated with the physical body (*sthula śarīra*).

Prāṇamaya is the vital sheath where life and its vital forces are experienced. Experiences such as breathing, hunger and the like are characteristic of this layer. This is the first subtle state.

Manonmaya is the sheath provided by the mind. Experiences such as thinking, remembering, being uncertain and the like are characteristic of this layer which constitutes the second subtle state.

Vijñānamaya is the sheath of the intellect. The experience of knowing is very characteristic of this layer.

The three more subtle states of *prānamaya*, *manonmaya* and *vijñānamaya* are included in the dream state. Together they are called the subtle body (*sūkṣma śarīra*).

Ānandamaya is the sheath of bliss. The experience of being happy is characteristic of this layer. Since this universal experience comes after deep sleep, it is correlated with that state of consciousness. This is called the causal body (*kāraṇa sarīra*).

We have earlier defined the concept of *jīva* which is the lower self. The essential characteristics of *jīva* are existence and consciousness, except that they are severely constrained by the five sheaths. *Jīva* establishes its identity with *Īśvara* (personal God) whose intrinsic characteristics are existence and consciousness without any constraints. *Īśvara* is the equivalent of *Brahman* in the waking state of consciousness which makes Him accessible to the devotees. We think of *Īśvara* as the creator with all the infinite qualities of omniscience and omnipotence. The constraints on *jīva* as enumerated in the model of the five sheaths are directly traced to spiritual ignorance. The suggestion, therefore, is that when the constraints are removed, that is, when spiritual ignorance is destroyed, the identity will become self-evident. Consequently, removal of these five sheaths is the main purpose of spiritual practice.

3.9 AUM (OM)

The three fields of thinking, speaking and acting are central to our worldly transactions. The usual notion is that thinking is the basis for speaking and that speech is essential to communication. In fact, there are many who have tread the spiritual path who would attest to the inexplicable experience of having received direct communion from their masters without indulging in any verbal exchange. One is reminded in this context of the beautiful devotional hymn *Śrī Dakṣiṇāmūrtistotram* by Śaṁkarācārya which tells of Lord *Dakṣiṇāmūrti* seated under a pipal tree, instructing his disciples in the serenity of total silence.

Metaphysical theory points out that the real basis for thinking is in the Being. Since thinking is necessary for speaking, there is also a more fundamental relationship that exists between Being and speaking. What we normally mean by speech is the *vaikharī* stage, whereas there are, in fact, three stages preceding it. The four stages, from the most to the least subtle, are *para*, *paśyantī*, *madhyama*, and *vaikharī*. The first three stages are supposed to be stationed in the middle of the body. The field of speaking, which is the *vaikharī* stage, is therefore grounded in the Being itself. Modern day linguists also inform us that there are deeper levels of structure associated with language. But the Being is beyond the realm of name and form, and it is inconceivable at first to think that any sound could invoke correlations with that mystical entity. It is here that the scriptures come to our assistance and point towards the existence of precisely such a primal sound of the universe.

The most sacred sound in the Vedic literature is the sound produced by three syllable word AUM. In the Sanskrit alphabet, 'A' is the first vowel and letter and so is the very first sound a man can utter. The pronunciation of 'M' involves the closing of the lips and is the last sound a man can utter. 'U' is the sound produced by rolling the breath over the whole of the tongue. Consequently, AUM pronounced as OM is a combination of all sounds that man can utter and so it has a universality about it. To quote Swami Ranganāthānanda from his book on *The Message Of The Upaniṣads* [36]:

> Om in its uttered form finally merges into its unuttered form; all

> uttered sound merges into the silence of the soundless. This soundless or *amātra* aspect of Om is the symbol of *Brahman* in Its transcendental aspect, beyond time, space and causality. This *amātra* aspect is indicated by the *bindu* or dot in the crescent over the syllable Om as written in Sanskrit.
>
> This Om, as the unity of all sound to which all matter and energy are reduced in their primordial form, is a fit symbol for *Ātman* or *Brahman*, which is the unity of all existence. These, and possibly other, considerations led the Vedic sages to accord to Om the highest divine reverence and worship , and treat it as the holiest *pratīka*, symbol, of divinity; they called it *nāda Brahman* or *śabda Brahman*, *Brahman* in the form of sound. It is the holiest word for all the religions emanating from India–Hinduism, Buddhism, Jainism, and Sikhism. Its nearest equivalent in the West is the Logos or the Word. As St. John's Gospel majestically expounds it: 'In the beginning was the Word, and the Word was with God, and the Word was God.'

It is considered to be the primal sound of the universe and is called the *vedavākya*. It is also called the *praṇava* and its repetition in some prescribed forms has the effect of taking the one who is chanting from the gross to the subtle levels of his being.

The following discussion presents some new terminology for the gross, subtle and the causal bodies and establishes their correspondence with the primal sound AUM. It leads to the conclusion that AUM is the experiencer in all the three states of consciousness and that beyond it is the universal consciousness. The *ātman* in identification with the gross body is called *virāt*. *Virāt* coupled with the waking state is called the *viṣva* state and is identified by the letter A of AUM.

The subtle body can be defined as consisting of the seventeen elements (five vital airs to be defined in the next section, five sensory organs, five motor organs, the mind and the intellect) and is called the *Hiraṇyagarbha*. *Hiraṇyagarbha* or Cosmic Person is the personification of the infinite consciousness. It is what a man, in his finititude, can visualize about the infinite. All human beings possess varying degrees of consciousness depending on the constraints imposed by their levels of spiritual ignorance, but the Cosmic Person is without any limiting adjunct and is therefore invested with the same infinite consciousness as the impersonal *Brahman*. Cosmic Person combines both the essential attributes of the personal and the universal. Personal consciousness is limited by space, time and causality whereas universality has the element of transcendence. Vedanta visualizes the Cosmic Person as the impersonal–personal God which is the link between the fields of being and becoming. This concept of God is different from that of monotheistic religions. In the Hindu view the universe also is made sacred by God's all pervasive presence. The *ātman* in identification with *Hiraṇyagarbha* and the dream state is called the *taijasa* and is denoted by the second syllable U of the *praṇava.*

The causal body of the *ātman* is *avyaktakara,* which means undifferentiated. It is the cause of the gross and the subtle bodies. The causal body is dispelled only upon final realization of the *ātman.* In the waking state, knowledge is obtained by the use of five senses. In the dream state, the sense organs are not functioning and consciousness appears both as the subject and the object; also one has some knowledge arising from the impressions of the waking state. In deep sleep, there is no duality of subject and object, and the determinative intellect will be in the causal condition. *Prajñā,* which is awareness, is defined as *ātman* in identification with the causal body and the deep sleep state. The syllable M of AUM corresponds to the causal body, the deep sleep state and the *prajñā.*

In summary, the experiencer in the waking state, *viśva,* is denoted by the syllable (A); the experiencer in the dream state, *taijasa,* is denoted by the syllable (U); and the experiencer in the sleep state, *prajñā,* is denoted by the syllable (M) of the primal sound (AUM). All these experiences merge into the pure consciousness. That is why the repetition of this sacred sound becomes a means of spiritual elevation.

3.10 Īśvara

The *Upaniṣadic* truth is that *Brahman* is the ultimate reality. Since we are living in the world of duality, we are naturally interested in matters pertaining to the world and also in the philosophical question of creation of this universe. Although *Brahman* is deemed as the ultimate truth behind the universe, we seek for its equivalent concept in a personal God, *Īśvara,* who is the creator and protector, since this more human concept bestows on us the hope that there is a definite possibility of reaching out to him even while we are fettered within the confines of our mortal limitations of finititude. We invest *Īśvara* with all the intrinsic qualities of *Brahman* which are existence, consciousness and bliss, and further ascribe to Him other divine qualities through which we can easily relate with Him. Thus what seems like the inaccessible transcendent is rendered more accessible through supreme love and devotion.

The Vedantic concept of *Īśvara* is very different from the understanding of creation in terms of man, nature and God above, as in the monotheistic religions. In these religions, God, who is vested with infinite knowledge and infinite power, is considered as the divine force whose abode is beyond the universe. Only His transcendental aspect is retained, not the aspect of His immanence. Monotheistic religions consider God as only the efficient cause, that is, the cause of creation of this universe. The material cause of the universe does not enter into the picture. He is supposed to have created this universe *ex nihilio.* In contrast to this thinking, *Īśvara* is considered as both the efficient and the material cause. To illustrate this idea, the usual example that is given is that of a spider weaving a web. The spider in this process is the efficient cause because it weaves the web and it is also the material cause because the material for the web also comes from the spider.

Vedantic thesis expects us to consider the *Brahman* of its philosophy and *Īśvara* of its religion only as a working hypothesis till the doubts about their validity are dispelled through spiritual practice. The ultimate aim, as always, is the concrete realization of these concepts by the spiritual aspirant. If this emphasis were not there, all these ideas would remain in the sphere of speculation and theological dogma.

The entire universe is considered as *Īśvara*'s cosmic vesture. Consequently, every discrete entity found in the universe is viewed as a fragment of His cosmic vesture. This immense diversity renders it possible to worship Him in a multitude of ways, a concept which is central to the theistic interpretation. One God is worshipped in several forms, and not several Gods, as is the case in pantheism. Because of the varied forms of worship, it is important to reiterate that Hindus believe in one, and only one, God. *Īśvara* is the universal consciousness from the vantage point of our own world of duality where we cannot avoid thoughts about the creator. But it is the immanence of the *Upaniṣadic* God that provides multiple ways of making Him accessible to mankind.

3.11 Meditation (*Dhyāna*)

Meditation (*dhyāna*) is the principal means employed for achieving purification of the mind so that it can progressively acquire the capacity to dissolve the effects of spiritual ignorance (*avidyā*) enveloping knowledge of the Self. While we have stated the purpose in terms of the concepts that we have developed in connection with the discussion of Vedic philosophy, it should not be overlooked that every religion has its own favorite form of meditation and has developed its own special vocabulary in philosophical terms to explain its purpose. Furthermore, there are a variety of meditative techniques available even within a particular religion. But one factor that is invariant within all these seemingly different disciplines is the recognition that meditation ranks high in the list of preparatory steps for gaining either spiritual knowledge or, equivalently, religious experience.

The technique of meditation is best learnt from a preceptor because it belongs to the realm of practice; however, a brief description of the theory behind it is useful. First of all, meditation requires something to focus the practitioner's attention. This is usually the name of a holy sound, a *mantra*, which could be in one or more syllables. Every tradition has its own set of *mantras*, and the choice of a particular one is exercised by the preceptor depending upon the level of preparation of the seeker. The *Karma* doctrine informs us that not everybody is born equal because of the different burdens that we carry as a result of the cumulative impressions (*saṁskāras*) gathered from our thoughts and deeds of our past lives. Consequently, the initial conditions for spiritual advancement vary from one person to another. The equality of man, on the other hand, is assured in terms of the realization that everyone is endowed with the element of infinite consciousness, which, in turn, ensures that every one is entitled to gain spiritual realization, albeit with varying degrees of preparation (*adhikārabheda*). The teacher is sensitive to such variations before launching a seeker on a spiritual path.

Apart from the *mantras* which are specific to the seekers, called *guru mantras*, there are many *mantras* that are in the public domain, such as the famous *gayatrī mantra* of the Hindus, the *mantra* of the Hare Krishna movement, and some of the *mantras* that were chanted aloud by some of the saints of the Hindu tradition. And finally, of course, there is the most famous holy word of Hinduism, the word OM (*praṇava*), which can be used as a *mantra*. Seekers are advised, however, to start meditation after going through a proper initiation ceremony under the guidance of a qualified teacher. A *mantra* is chanted mentally in total silence after assuming a special sitting posture with eyes closed. One is advised to meditate at a regular time every day for a period of twenty to thirty minutes. The duration may be extended in a very natural way by an experienced meditator.

The process of meditation consists of repeating one's *mantra* within the serenity of one's own inner silence in a most relaxed and effortless way. The idea behind repetition is that it will have the effect of slowing down the activity of a meandering mind which goes from one thought to another in an uncontrolled manner. A wandering mind has the natural propensity to follow the ordinary processes of life which implies that it always projects outwardly to the external world. In order to reorient it towards the inner Self, which in fact, is its true abode, its nature of wandering has to be curbed. Since the mind is nothing but a stream of thoughts, the silence that resides between two thoughts is missed by a restless mind. The restlessness is due to the natural tendency of the mind to follow the objects identified by successive thoughts. When these objects are different from one another, as is often the case, the mind keeps wandering, thus missing the silence that resides between two thoughts. The mechanism for keeping the unbridled mind under control consists of breaking the relation that exists between two successive thoughts.

The repetition of the same holy word, which is the principal mechanism of meditation, enables one to experience the silence that exists between two successive thoughts. In the absence of a relation between two successive thoughts, the waking state of the mind gets imprisoned between two similar thoughts which is the abode of total silence. With some practice, the thoughts get progressively feeble until they vanish altogether resulting in experiences of total silence for longer intervals of time. The absence of experience of 'flow of time' during meditation stands in contrast to its experience when one is fully in the waking state. The experience of the 'junction point' between the repetition of two successive chanting of the *mantra*, where there is only silence and no thoughts, is also called the experience of the transcendental field which is above our ordinary three states of consciousness.

The experience of transcendence will not result overnight, at least not in the majority of cases, because there will always be the tendency for the mind to escape the closed loop of two successive thoughts of a *mantra* and, instead, start experiencing stray thoughts. When this happens, as is usually the case, the meditator is advised not to attach any importance to those stray thoughts but to slowly return to the focus of his meditation which is his *mantra*. With practice the wandering mind loses its propensity, and the frequency of such interruptions will also lessen. The point that is made is that the nature of wandering is not the intrinsic characteristic of the mind, but it appears so because at any given moment it is incessantly in search of a higher field of happiness. When the mind discovers that the experience of the transcendental field is the true abode of happiness, it loses its penchant for wandering. A restless mind is comparable to a honey bee which quietly settles down on a honeycomb after indulging in aimless wandering. With constant practice, it is possible to attain a stage where one gets firmly entrenched in the higher state of consciousness (*samādhi*).

The Hindu tradition recommends that the stage of meditation be preceded by two stages: first a prayerful worship (*pūjā*), and secondly oral chanting of *mantras* (*japa*). Completion of these preparatory stages, which are also meant for the purification of the mind, is believed to assist in successful practice of meditation (*dhyāna*) which is nothing but *japa* that is conducted in total silence according to the method discussed earlier.

3.12 Māyā

Earlier we came across the concept of *māyā* (illusion) which we shall discuss at some length because of its importance in the philosophy of nondualism. *Māyā* has two distinct aspects to it.

- It conceals the nondual nature of *ātman*.
- It projects the external world in all its diverse aspects.

We use two different words to connote the two distinct aspects. The word *māyā* is used when we want to emphasize the first aspect, whereas the word *avidyā*, meaning spiritual ignorance, is used to emphasize the second aspect. It is possible for a person to be extremely bright in worldly matters and yet at the same time to be spiritually ignorant. The classical metaphor for *avidyā* is the darkness experienced by an owl in broad day light.

Māyā does not mean that the world that we are living in is an illusion as is commonly misinterpreted. Rather, the concept of *māyā* arises from our mistaken understanding that one can arrive at the absolute truth from an investigation of worldly phenomena as in scientific endeavors. The terminology can perhaps be best understood with reference to the development of theoretical models that we encounter in scientific studies. For instance, a physical theory, whether it be the Newtonian theory of classical mechanics or the relativity theory due to Einstein, is recognized as only a representation of reality rather than reality itself. As the semanticist Alfred Korzbski put it, the map is not the territory. It is so very easy to commit the mistake of confusing the map with the territory, or the model with the aspect of reality that it represents, in which case it becomes an illusion. This idea is further elaborated on in the following quotation from Aldous Huxley's book on *Ends and Means* [24]:

> From the world we actually live in, the world that is given by our senses, our intuitions of beauty and goodness, our emotions and impulses, our moods and sentiments, the man of science abstracts a simplified private universe of things possessing only those qualities which used to be called 'primary'... Arbitrarily, because it happens to be convenient; because his methods do not allow him to deal with the immense

> complexity of reality, he selects from the whole of experience only those elements which can be weighed, measured, numbered, or which lend themselves in any other way to mathematical treatment... The success was intoxicating and, with an illogicality which, in the circumstances, was doubtless pardonable, many scientists and philosophers came to imagine that this useful abstraction from reality was reality itself. Reality as actually experienced contains intuitions of value and significance, contains love, beauty, mystical ecstasy, intimations of godhead.

The central purpose of *Vedāntic* inquiry into truth is to articulate ways and means for dispelling this confusion for ever.

Māyāvi is the wielder of *māyā* when the latter is likened to a *śakti*, a force which acts as a veil between the waking state of consciousness and the pure consciousness. This force is analogous to the force one experiences while going to sleep (*nidra-śakti*), which takes one from the waking state to the sleeping state of consciousness.

We can deduce several conclusions based on the concept of *māyā* and its twin facets. It is *māyā* that gives us the feeling that we are finite entities, which in turn, leads us to the perception that the universe is external to us. The limitation one experiences with respect to what one can know or do can also be explained on the basis of the constraint imposed by *māyā* on the individual. The positive side to this feeling of separateness from the external world is that it has given rise to the scientific endeavour to understand nature in all its facets even when we know that we cannot arrive at the ultimate reality through such a pursuit. But it seems as though that the universe needs the consciousness of man, limited though it might be, to reveal its own infinite glory.

Although pure consciousness is nondual in nature, the individual mind faces obstacles in the way of realizing this truth. These obstacles are attributed to *māyā*. The whole purpose of spiritual practice is to reduce the veiling influence of *māyā*. Spiritual knowledge is very different from secular knowledge. The knowledge of Self is not considered as new knowledge since it is already there, though it is veiled from experience because it is masked by spiritual ignorance. Secular knowledge, however, is not already present, and so effort is required to learn new facts and concepts. For instance, the efforts that are required to know something new about cosmological models of physics are of a totally different kind than those required for knowing one's own Self which is already present. Basically, it is not even an intellectual effort. It can only be realized on the basis of actual experience although an earlier intellectual awareness of it is considered to be extremely helpful.

Māyā does not have an independent existence apart from the abiding universal consciousness. It may seem independent because it has the power to present a distorted view of the Self. It can cause serious problems because of its capacity to present itself as something independent of the universal consciousness.

Māyā is a phenomenon which can be viewed as being both present as well as absent. When viewed from the coordinates of empirical reality, it is existent because one can experience its diversifying power. On the other hand, when it is visualized from the platform of nondual consciousness, there is neither concealment nor diversification and therefore it is non-existent. There can be no proof for the concept of māyā since māyā also is a non-existent. A *pramāṇa* can only be invoked to remove the ignorance concerning an existent. It is only ignorance concerning an existent that can be removed by a *pramāṇa.*

Avidyā, which is spiritual ignorance, is considered as without a beginning since its locus encompasses the Self which is eternal. It is considered to be the result of mutual interaction between three of its constituents which are called *guṇas.* These are *sattva, rajas* and *tamas. Sattva* refers to the aspect of purity, *rajas* to the quality of action, and *tamas* to the trait of inertness. These are the basic constituents of nature which are responsible for its evolution which, in turn, presupposes the concepts of creation and progressive development. In the psycho-physical mechanism of an individual, these qualities appear in unique proportions. Two individuals differ with respect to the proportions of these three constituents. The practice for overcoming spiritual ignorance consists of progressively increasing the *sattva* content of one's personality makeup.

The power of cognition is a mode of *māyā* and is called the mind. All spiritual ignorance is inextricably linked to the mind, since it is this faculty which is responsible for maintaining the illusory difference between the soul (*Jīva*) and the *Brahman.*

Chapter 4

Science & Vedic Philosophy: Bridging Concepts

4.1 The Fourth State of Consciousness

The psychic principle, which is the immanent reality that is identical to the transcendental reality, is also known by several other names in the philosophical literature. Pure consciousness (*turiya*), plenary consciousness, fourth state of consciousness, Self with a capital S, and eternal witness are some of the other terminologies that are used very frequently. Although we have said that the existence of the Self can only be experienced and is beyond the realm of intellectual understanding, we find nevertheless various attempts to draw our attention to its possible existence. These serve only as pointers to the unmanifest field of existence from the plane of the manifest worldly realities. It is a pointer in the sense that while pointing to the moon we say that the moon is at the tip of the branch. The assumption here is, of course, that the unmanifest reality leaves behind some traces which can be detected in our ordinary worldly consciousness if we pay careful attention to them.

Since Hindu metaphysics proceeds with the investigation of the ultimate reality from the level of the individual self, analysis of the three states of consciousness, namely the waking, the dreaming and the sleeping states, assumes paramount importance. The search is for the real 'I', which is the feeling of constancy in the field of being. *Who am I* is the real question to be answered. Whenever we seek an answer to this philosophical question proceeding from a careful investigation of some global features of the three states of consciousness, the inquiry sometimes gets wrongly classified as psychological analysis. It is therefore important to state at the very outset that this type of analysis of the three states of consciousness does not rely on the subject of modern-day psychology. Furthermore, the origin for such analysis goes back to antiquity, to the *Upaniṣadic* era. We are only interested in probing the question of existence from the vantage point of the observer and the observed in order to get a clue as to the true nature of constancy of the real 'I' that exists in all the triad of states.

The analysis of the subject and object relationships in the three states of consciousness comprehends the totality of experience, unlike scientific analysis which is confined only to the data of the waking state (*jāgrat*). The search for the core meaning of existence when directed towards our inner universe instead of the external world necessitates taking the totality of human existence into its domain of investigation. The *Māndukya Upaniṣad* deals with a detailed study of the triad of states and points to the existence of yet another state of consciousness called *turiya*, the pure consciousness, which is at once eternal and non-dual in nature. The conclusion is that this fourth state of consciousness coexists with all the other three states and is therefore the real 'I' which is the constant factor in the paradoxical conjunction of the being and becoming. That it is also the real ground for the realm of becoming is known from our discussion of the Vedic philosophy. The following observations are intended as an elaboration of this central theme.

It is only in the waking state that one experiences the three principal features of the external world, namely, space, time and causality. Also, these categories of thought are never sublated in that state. We can refer to these as a totality of requisites that are essential for a complete manifestation of the external world with all its diversity. That is, it is only in this state that one can experience the diversity of the universe, and also, intellectually comprehend that one is veiled from the experience of the ultimate reality. In the dream state, on the other hand, the totality of requisites are completely distorted and, therefore, are considered to be absent. For instance, one can dream about a long interval of time within a short duration or perform superhuman feats like jumping over long distances. The experiences of the waking and dream states are quite distinct, although in both, the duality of the observer and observed relationship, namely, the subject and object relationship prevails. However, the conscious state and the dream state are mutually exclusive. It is in our experience that as soon as we wake up, the reality of the dream state gets instantly snuffed out. On careful examination we find, however, that it is the reality of the ordinary waking consciousness that refutes the experience of the dream state when that memory is retrieved in the waking state; it is therefore tantamount to one waking experience refuting another waking experience. The refutation always takes place only in the waking state.

It is all too common to refer to the experiences of the three states of consciousness as the realities corresponding to those states. However, it is important for our thesis to note that it is only a loose reference for convenience in communication only. Neither of the experiences of the waking or dream states correspond to the ultimate metaphysical reality because of their ephemeral nature. Reality, by definition, is something that can never be sublated. It is only the experience of the psychic principle, the Self, that belongs to that category. The realities of the waking and dream states do not satisfy this definition since they can both be sublated. Consequently, the use of the word 'reality' in reference to the two states should be understood only in the relative sense.

In the state of deep sleep (*suṣupti*), there is absolutely no experience of duality, and hence it is not a state in the ordinary sense of the word. There is no experience of either the subject or the object. One thing that is characteristic of this state while in it, is that one does not experience the passage of time, unlike being in the other two states. It is indeed a non-dual experience that is characterized by total bliss which can be recalled only when one wakes up. It does not follow, however, that the Self is perceived in the sleeping state. For that to happen, it has to be first perceived in the waking state because that is our true nature with respect to the triad of states. The bliss that one enjoys after waking up from deep sleep is an universal experience and is indicative of the intrinsic property of bliss, *ānanda*, which is one of the three epithets in *saccidānanda*, that we had used to describe *ātman*. The absence of cognition in deep sleep is attributed to the absence of any object of perception and is not due to the absence of consciousness since the eternal witness is always present and coexists with the sleeping state also. The collapse of the observer– observed relationship means that there is nothing separated from Self due to spiritual ignorance.

The fourth state of consciousness, which is the pure consciousness, is also referred to as a state simply because we have referred to the other three as states. Strictly speaking, it should not be referred to as a state since it is something that is ever present and beyond the requisites of space, time and causality. Pure consciousness coexists with the other three states and is therefore referred to as the eternal witness, *sākṣi*.

The relationship of the fourth state of consciousness with respect to the conscious state is indirectly inferred on the basis of the analogous experience of the waking state with respect to the dream state. Incredible as it might seem, this analogy does provide a conjectural insight. The conscious state is where we perceive the reality of the world and that is the state associated with all our scientific investigations. One could now imagine the possibility that we 'wake up' from the 'dream' of the physical world to switch on to the fourth state of consciousness. From the platform of the fourth state, the 'I' of the waking state would have an illusory existence just as the 'I', the dreamer, would cease to exist in the conscious state.

Computer technology, specifically the concept of virtual reality, provides some further insights into the understanding of the preceding discussion. Practically every phenomenon that is within the scientific purview can be simulated on the digital computer. It could be a simulation pertaining to the landing on the moon or a simulation of the DNA sequence of molecular biology. It is possible to envisage a day when human consciousness can also be simulated. In such a case, the entities with consciousness inside the computer would behave as if they had dominion over the world of simulation and would not be aware of the fact that they are being controlled from outside by a human programmer. Extending this analogy, one can perceive the entire universe as a gigantic computer wherein we, the individual human beings in the waking state, are in a world of simulation impervious to knowing that we are being controlled from outside. The phenomenon of virtual reality, which for the present is only a concept, provides an analogy for the relationship of the dream state to the waking state of consciousness to explain the illusory notion of the waking state viewed from the vantage point of pure consciousness.

4.2 A Body-Mind Relationship

Direct instruction for the realization of *ātman* is impossible because it is beyond the reach of the *pramāṇas* (valid means of inquiry) of rational knowledge. It can only be accomplished by indirect means through the exercise of skill coupled with a knowledge of the level of spiritual preparation of the seeker. Spiritual practice is meant to gradually lessen the veiling influence of the constraints and weaken the hold of the limiting adjuncts (*upādhis*), which mask the Self from direct experience. The only means at our disposal is to design ways and means for reorienting the very constraints in the direction of unveiling their influence. For instance, if the natural propensity of thought is to dart out to the external world, we have to design methods by which it can project inwards into the deep recesses of the mind. The subtler the constraints, the more powerful they are. The two subtlest candidates singled out for spiritual practice are the mind (*prajñā*) and the vital airs (*prāṇa*) as exemplified by breathing. The high degree of their subtlety is determined by their close proximity to the sentient *ātman*. The body ceases to function in the absence of *prajñā* and *prāṇa*, a fact which again reinforces the observation. Thus, mind, which is responsible for cognition, serves as an important gateway for the realization of the Self. Just as it has the capacity to identify itself with the physical body, it is equally capable of shifting its focus to the Self, and hence its importance. Figuratively speaking, it can be referred to as an amphibian because of its dual capacity to reach the opposite banks of a human personality, or its two antipodes, with equal ease. The closely associated and interdependent constraint, *prāṇa*, is responsible for action. That *prajñā* and *prāṇa* are interdependent can be observed from the experience that when one slows down the rate of breathing as in *prāṇāyama*, a yogic discipline, the mind also calms down. Conversely, when the mind is calm, breathing also slows down. Next, we discuss some further properties of the mind and the vital airs in order to shed more light on their importance in spiritual practice.

The mind can undergo four types of transformations, called *vṛttis*. The first one is *manas* which is the transformation responsible for weighing the pros and cons of a thing. Second is *buddhi* which is the mental faculty required for determining the true nature of objects, the *jñānaśakti*. *Ātman* with the limiting adjunct of *buddhi* is called the cognizer and it is this entity that is responsible for the three states perception, non-perception and misperception. The third type of transformation of the mind is *ahaṁkāra* which is the Ego factor responsible for identifying the body with the Self. The usual interpretation of a philosophy based on materialism is based on the ego factor. Lastly, *citta* is the faculty responsible for remembering things of interest. The mind operates through the five senses and their corresponding five motor organs which are called *indriyas*.

We now list the five functions of the primary vital air, *mukhyaprāṇa*, associated with the human body, which is instrumental for its role as a doer, *kriyāśakti*; *ātman* with the limiting adjunct of the vital airs is called the doer. *Prāṇa* is associated with the function of exhaling; *apāna*, of inhaling; *vyāna*, that which exists between inhaling and exhaling; *udāna*, the vital air observed during such acts as departure from the body; and finally, *samana* which is responsible for carrying the essence of food to all the limbs.

Yet another definition of the subtle body is provided on the basis of this analysis which includes the concept of *prāṇa*. The subtle body consists of five motor organs; five sense organs; five *prāṇas*; five elements which are space, air, fire, water and the earth; the mind in its four aspects; and *avidyā*, desire and *karma*. As always, the entire range from the gross to the subtle proceeds in successive layers. At the outermost periphery are the physical aspects of man, and as one penetrates into the innermost layers one finds layers in an ascending order of subtlety, immensity and fineness.

The subtle body is also called *Linga-śarīra*, and it consists of all our latent impressions acquired as a result of our actions over the long history of the soul. One could say that this is the cognitive burden that a man carries when he perceives the world around him since his perception is always colored by the traces that are left behind by his past experiences over many lives. Consequently, *Linga-śarīra* acts as a constraint, a limiting adjunct (*upādhi*), on the Self.

4.3 Some Medical Findings About Consciousness

During the past few decades, scientists have succeeded in establishing definite physiological and biochemical correlation with the three states of consciousness and continued with their investigations into the possibility of coming up with such correlates for the fourth state of consciousness also. There is evidence beyond doubt that the waking, dreaming and deep sleep states of consciousness have corresponding physical states in the human nervous system. Of these, the correspondences of the waking and sleeping states were discovered first, and the correlation for the dreaming state came a little later. Some of the measurements made are the metabolic rate, heart rate, cardiac output, Galvanic skin resistance (GSR), electrocardiograph (EKG), electroencephalograph (EEG), rates of oxygen consumption and carbon-dioxide elimination, and concentration of blood lactate. Detailed records of these measurements showing their unique correspondences with the three states are available in the scientific literature of many countries.

Perhaps the most extensive investigations to discover possible correlation with the fourth state of consciousness were made in the early seventies by Robert Keith Wallace, who is actively associated with Maharishi Mahesh Yogi of the transcendental meditation (TM) movement and Herbert Benson of the Harvard Medical School. Undoubtedly, their sample of meditators was acceptable by any scientific standards because it included, in substantial numbers, people of all age groups belonging to both sexes and from several countries. Furthermore, the sample included people who had practiced TM for varying periods of time and therefore ranged from new initiates to well-established practitioners of the technique. It was possible to assemble such a sample because the technique of TM had taken root in several countries, and the technique itself is claimed to be a simple one to learn. As such the investigation did not suffer from difficulties that prior researchers faced in the daunting task of proceeding with a scientifically meaningful sample. The definitive paper on the research findings by Wallace and Benson appeared in *Scientific American* [47] and was an eye-opener to all those who were interested in the subject. The results were corroborated later by other research groups of several countries and published in prestigious journals. Based on these findings, one can definitely come to the conclusion that the state produced by TM is quite distinct from those encountered in the other three states of consciousness. They are also different from such other altered states that one finds in hypnosis, autosuggestion etc.

Broadly speaking, all the physiological and biochemical correlates of the TM state indicate a state of deep rest and relaxation in an order of magnitude greater than one finds in the deep sleep state. Maharishi Mahesh Yogi has called this state, a state of 'restful alertness' because of the simultaneous presence of a deeply resting body and a mind which is fully awake. He points out that it is in this 'wakeful hypo-metabolic state' that the nervous system is in its most healthy state which constitutes the basis for all energy and action. The medical benefits of TM have been studied in great detail and recorded in the published literature of the movement. Those who express a wish to get initiated into the practice of this type of meditation are first told about the research findings on TM and the plethora of advantages accruing from the meditative discipline, and then gradually led on to the spiritual practice which is the ultimate aim of the movement.

There is a word of caution that has to be interjected at this stage. We have characterized pure consciousness as a non-sensory experience and asserted that only the scriptures can bear testimony to it. Instrumentation which is at the basis of the findings of the physiological correlates of consciousness come within the category of sensory data, and as such we have to be careful about admitting these experimental results as the validation for the existence of a fourth state of consciousness. Maharishi Mahesh Yogi, himself an ardent devotee of the Śaṁkarācārya tradition and a physicist by his early training would not, of course, commit such a conceptual mistake. Careful investigation will lead to the conclusion that the results, impressive as they are, serve only as pointers to the fact that the human personality is not confined only to the triad of states. The basic fact is that even the consciousness of the waking state is also a causal reality, and as such it does not send out any physical signals. It is something that is understood only on the basis of experience. The real value of these scientific findings is that they have the demonstrated potential of impressing on skeptics that there is more to the human personality than they had imagined and convincing them that there is a tremendous source of inner strength waiting to be tapped for greater enrichment.

In this connection, Herbert Benson's book on *The Relaxation Response* is also of great interest. The relaxation technique which is based on meditation is now routinely suggested to patients who have suffered heart attacks. Benson, who in his early career was associated with Wallace in the experiments on TM, has taken pains to develop an independent thesis suggesting that transcendental experience is not unique to a particular religion and is fairly widespread in all faiths. Although this is by no means a novel claim, he has perhaps broadened the appeal for his technique so that conscientious objectors to the practice associated with any one particular religion can also benefit from the suggested treatment. To quote Benson [3]:

> The relaxation response is a natural gift that anyone can turn on and use. By bridging the traditional gaps between psychology, physiology, medicine, and history, we have established that the relaxation response is an innate mechanism within us.

Benson's sole focus of his meditative technique is to induce relaxation in order to relieve accumulated stress. He has carefully avoided all references to meditation as a spiritual practice, although that was unquestionably its original purpose. Perhaps the deeper significance of Benson's relaxation technique as well as the research on TM is the clear establishment of a body-mind relationship that was scrupulously avoided in the purely mechanistic view of the world heralded by classical physics.

4.4 Several meanings of Reality

Since the philosophical inquiry of *Vedānta* is very much concerned with probing into different experiences of reality, we will discuss these ideas in some more detail. The speculation about the ultimate reality associated with the universe has long engaged the attention of philosophers throughout the centuries. It is broadly discussed under two headings, the objective reality and the subjective reality. There are those who argue that the universe is something that is 'out there', a stark objective reality, in the sense that Mount Everest exists irrespective of the observer. And there are others who believe that the perceptions of reality of the universe that we entertain are our own mental constructions which are subjective in nature, quite independent of any reference to objective reality. We shall briefly dwell on these contrary views because of their relevance to the *Vedāntic* inquiry about reality which is quite distinct from either of these two theses.

We have already dealt with objective reality in science in chapter 1 while commenting on the mechanistic view of the world. The fundamental assumption is that the study of the external universe does not depend on the nature of the observer. In fact, in a certain sense, no scientific study would have been possible without the explicit recognition of the different entities in the universe and without the specificity suggested by their names (*nāma*) and forms (*rūpa*). We also commented on certain branches of mathematics which are more in the nature of discoveries rather than inventions. The very fact that much scientific endeavor consists of investigation into the laws of nature presupposes that we assign an objective reality to the universe.

In opposition to this view, we have the subjective view of reality. In fact, this was the principal point made by Immanuel Kant in his 'Critique Of Pure Reason', about which we have the following quotation from Barrow and Tipler [1]:

> All our empirical inquiries into the structure of Nature regard it as an entity which incorporates within itself a system of empirical laws. These laws are unified and naturally adapted to the faculties of our own cognition. The design we perceive must be necessarily mind-imposed and subjective to our innate categories of thought. Although the 'the things in themselves' are mind-independent, our act of understanding completely creates the categories in terms of which we order them. Inevitably, we view the world through rose-colored spectacles. These self-created categories cannot themselves be ascertained by observation; they are *a priori* conditions of the experience we have, like the perception of the space-time continuum. We cannot through our experience hope to ascertain the conditions of such experience. Our observation of order and structure in the universe, he argues, arises inevitably because we have introduced such concepts into our analysis of experience.

It is interesting to note that even in scientific investigations, the philosophically inclined scientists are also looking for ever-increasing orders of reality which take both the objective and subjective views into account. We refer to the work of David Bohm, a renowned British physicist, to illustrate this idea further. His book on *Wholeness and the Implicit order* [9] approaches the problem from the vantage point of theoretical physics. Bohm, in addition to his expertise in quantum mechanics, had extensive contacts with the late Indian philosopher J.Krishnamurthy. Although he does not explicitly mention that he is dealing with the question of mysticism, it is quite apparent that he is thoroughly probing that very problem from the vantage point of science. The following quotation from his book [9] makes interesting reading:

> Man's first realization that he was not identical with nature was also a crucial step, because it made possible a kind of autonomy in his thinking, which allowed him to go beyond the immediately given limits of nature, first in his imagination and ultimately in his practical work.
>
> Nevertheless, this sort of ability of man to separate himself from the environment and to divide and apportion things ultimately led to a wide range of negative and destructive results, because man lost awareness of what he was doing and thus extend the process of division beyond the limits within which it works properly. In essence, the process of division is a way of *thinking about things* that is convenient and useful mainly in the domain of practical, technical and functional activities. ... Being guided by a fragmentary self-world view, man then acts in such a way as to try to break himself and the world up, so that all seems to correspond to his way of thinking. Man thus obtains an apparent proof of the correctness of his fragmentary self-world view though, of course, he overlooks the fact that it is himself, acting to his mode of thought, who has brought about the fragmentation that now seems to have an autonomous existence, independent of his will and of his desire.

Bohm's prescription, of course, is that man should focus on ways and means for putting an end to the habit of fragmentary thought. Only by doing so is there a possibility of apprehending the reality of nature as a whole, so that the response also will be whole.

Although no one has come up with a 'spiritual calculus' to describe the order corresponding to wholeness because of the utter impossibility of the task, Bohm has made a serious effort to come up with mathematical models to describe some *unobserved orders in nature.* He was led to this research in connection with the explanation of some paradoxes encountered in quantum mechanics which are usually glossed over by many scientists. Since he is also deeply involved in the problem of mysticism, he was interested in probing into the philosophical aspects implied by the paradoxes in quantum mechanics. To quote Josephson [42], a Nobel laureate who is well acquainted with both Bohm's work and the Indian meditative techniques with their underlying philosophy:

> What he proposed is that the fact that things don't seem to be well determined in science means that there are variables which we can't observe directly, which do, however, effect the physics. His views gradually

> evolved, and what he says now is that what these paradoxes show is the existence of an unmanifest or implicate order which we can't observe directly, and phenomena are created from this order rather in some sort of way in which a cloud condenses from moist air. We see order in the observed phenomena which is the result of the unobserved order... if we want to put God into science, then we have to say that there is an intelligence behind the scenes which is creating order or at least leaving things less disordered than they would have been without the intelligence being present. And so we can identify the unobserved order with intelligence. Here we are on the way to the beginnings of a mathematical synthesis.

Josephson goes on to establish some interesting correspondences between the three grades of physical reality and the experiences of man which include both the experience of the waking state and meditative experiences. Although he cautions us that these are speculative in nature, they are interesting nonetheless because of the scientist's constant search for such parallels. His conclusion is that the sensory experience of the waking state corresponds to the reality of classical physics; that the physical reality of quantum mechanics corresponds to the subtler realities of the astral and celestial worlds; and finally, that Bohm's unmanifest or implicate order corresponds to transcendental experience.

The research on mathematical models for unobserved orders in the universe is still an ongoing one. What is really interesting about it is the implication that as science progresses in its understanding of nature, it has to continually contend with aspects of reality that always elude the scientist's observation and for which he has to postulate the existence of unobserved orders.

We shall now return to the view expressed by *Vedānta* on the question of subjective and objective reality. One of its chief exponents, Śaṁkara presents a view of ultimate reality which takes both objective and subjective views into account in order to establish the truth about the Self. He assigns an objective reality not only to the entities which we normally consider as empirically real, like Mount Everest, but also to objects we consider illusory like a rope being mistaken for a snake. According to him, wrong realizations arise because of spiritual ignorance (*avidyā*). In this analysis, the wrong realizations are not attributed to erroneous subjective experiences, but to an improper cognition of the objective reality. This is the most crucial point of the analysis.

That the external universe is perceived differently by different people is also a basic insight of modern-day cultural anthropologists. It is their finding that people belonging to different cultures perceive different realities when exposed to the same complex experience. These wrong realizations are attributed by *Vedānta* to different modifications of *avidyā*, which is a facet of *māyā* operating at the individual level. The contradictions inherent in our experience of the world and in our knowledge of it as exemplified in the two views of objective and subjective reality will always remain at the sensate level. This fact is a direct result of the superimposition of the not-self on the Self. This riddle can never be solved unless we bring the Self or *ātman* into consideration. Through a detailed argument, what Śaṁkara establishes is that only when spiritual ignorance is completely eradicated is it possible to perceive the objective reality free of any error. For example, the notion of one world is a figment of our imagination. Because of our spiritual ignorance we can only perceive fragments of this one world and then only one part at a time. But the entity which we call Self, which is the undifferentiated consciousness that can be experienced when one is free from spiritual ignorance, is wholeness itself, and it is the true witness to any objective reality. We will enlarge the scope of this discussion by leading up to a view expressed by Swami Vivekānanda.

Vedānta poses the question of Reality in the much broader perspective of the totality of all existence. The Sanskrit word for world is *loka* which includes both what is seen as well as what is experienced. Furthermore, all beings who have awareness are brought into this enlarged framework for a discussion of reality. Even confining the discussion to the realities revealed by the senses, it can be seen that there are any number of these realities. For instance, the *loka* of a creature with fewer than five senses will be entirely different from the *loka* of a human being. The concept of *loka* therefore not only includes the objective physical universe of man but also the many universes experienced by all beings. There is an infinity of such accounts of reality experienced by the awareness of all beings. *Brahman*, the ultimate reality, includes all of them since nothing is outside its purview. One interesting observation that is made regarding this vision of reality is that there is a steady convergence of the renderings of reality as one gets close to the realization of *ātman*, whereas the opposite traverse is marked by an increase in divergence, which we recall is an aspect of *māyā*. We have a long quotation from Swami Vivekānanda as further elaboration of this important conclusion [36]:

> We see, we must find the universe which includes all universes. We must find something which, by itself, must be the material running through all these various planes of existence, whether we apprehend it through the senses or not. If we could possibly find something which we could know as the common property of the lower as well as of the higher worlds, then our problem would be solved. Even if by the sheer

force of logic alone we could understand that there must be one basis of all existence, then our problem might approach to some sort of solution. But this solution cannot be obtained only through the world we see and know, because it is only a partial view of the whole.

Our only hope then lies in penetrating deeper. The early thinkers discovered that the farther away they were from the center, the more marked were the variations and differentiation; and that the nearer they were to unity... We first, therefore, want to find somewhere a center from which, as it were, all the other planes of existence start, and standing there we should try to find a solution. This is the proposition. And where is that center? *It is within us.* The ancient sages penetrated deeper and deeper until they found that in the innermost core of the human soul is the center of the whole universe. All the planes gravitate to that one point. That is the common ground, and standing there alone can we find a common solution.

4.5 Pramāṇa

A *pramāṇa* is a proximate means to valid knowledge (*pramā*). The purpose of a *pramāṇa* is twofold: first, it is meant to verify what is already known, and secondly, it is meant to assist in the acquisition of new knowledge. All the six systems of Indian philosophy present their own critique of knowledge for which they have to rely on the *pramāṇas* that they are going to employ, thus according *pramāṇas* a central place in their discussion. In fact, one method of classifying a system of Indian philosophy is based on the pramanas that it makes use of.

There are six types of *pramāṇa* that are recognized in the Vedic literature: 1) Perception (*pratyakṣa*) 2) Inference (*anumāna*) 3) Comparison (*upamāna*) 4) Presumption (*arthapatti*) 5) Non-apprehension (*anupalabdhi*) and 6) Verbal testimony (*śabda*). We shall only comment on the three major *pramāṇas* of perception, inference, and verbal testimony. Much of the ensuing discussion is meant to shed light on why verbal testimony is accorded the separate status of an independent *pramāṇa*.

Normally, in perception, there are three entities involved: a subject who sees, an object that is seen and the process of knowing. There are various theories covering the mechanisms of perception. The one in which we are interested most is called the *theory of representative perception* which is acknowledged by more than one school of Indian philosophy. We shall deal with this when we take up the discussion of these schools. The most characteristic feature of direct perception is that it is instantaneous. That is why it is considered a non-dual experience: there is a merger of the subject and object when perception occurs. It is because of this that even the experience of the transcendental reality is referred to as 'seeing the Truth' in order to emphasize the aspect of direct perception rather than visual perception. In fact, in Sanskrit, the word for 'philosophy' is *Darśana* which means 'seeing'.

Inference is the testimony that we seek whenever direct perception is not possible because of inadequate data which results in uncertainty. In such cases, we devise indirect means for inferring the truth. One such oft-quoted example for this situation is the inference that there is fire when there is smoke. This method of ascertaining truth is covered in the highly sophisticated mathematical discipline of probability theory. The various methodologies for inference proceed from a knowledge of only samples of the total inaccessible data in order to arrive at the most probable solutions for problems which are characterized by uncertainty. Essentially, these methods are devised for minimizing uncertainty inherent in many problems governed by probability.

Analogical reasoning is also included in the discussion on inference. We know a situation A in all its details, and we find another situation B where we only know that there are some identities that can be established between B and A. From a knowledge of that correspondence, we can reason out the details of B based on the basis of our knowledge of A. This method of reasoning is easier understood on the basis of quantitative models that are employed in scientific studies. We come across any number of cases where similar mathematical models appear for the study of disparate systems. For example, by establishing the one-to-one correspondence between electrical networks and properties of a mechanical system on the basis of their similar mathematical descriptions, we can study the property of one from a previous knowledge of the second. Perhaps a better example is to point out the modern day instrumentation techniques where most physical measurements are taken using electrical variables implying the existence of analogies to establish such correspondences.

We next discuss the *pramāṇa* of verbal testimony which is our major focus of attention. Statements, whether written or spoken, from people of honesty and integrity, can be very significant when they are used, for instance, as evidence in the dispensation of justice. But the question can arise whether this type of evidence deserves the status of a *pramāṇa* with its own distinctive logical status. The assertion is that verbal testimony is entitled to such a consideration for establishing the truth of metaphysical phenomena.

Those who dismiss verbal testimony as a separate category insist on relying exclusively on the time–tested *pramāṇas* of perception and inference, that is, on the valid means of inquiry applicable to worldly realities only. This logical stand deliberately ignores hints about the possible existence of the transcendental realm because it is ordinarily inaccessible to us in our waking state of consciousness. But our curiosity for furthering our knowledge should tell us that we do not cry a halt to investigations begun on the strength of hints simply because the ultimate reality is beyond of our immediate grasp. This difficult situation legitimizes a *pramāṇa* which is applicable only to matters that are beyond the reach of perception and inference. Vedas constitute such a *pramāṇa* because it validates matters pertaining to the transcendental realm. The visions of the sages who have attained self-realization and whose intuitive knowledge has been passed on to mankind through tradition (*smṛti*) constitute the preliminary validation for this unique testimony. But the question that can be legitimately raised is whether we can totally depend upon human insight, even that of the great sages, for guaranteeing the validity of the transcendental reality. Furthermore, there are mystics who may not necessarily subscribe to any particular faith. Consequently, despite our immense regard to the intuitive wisdom of the sages, we cannot put the new testimony on a sound footing because we cannot escape the criticism of subjectivism if we take that route. On the other hand, because Veda is viewed as revealed knowledge emanating from God and hence eternal, it does not carry the stigma of subjectivism. In the ultimate analysis, we can only conclude that ultimate philosophical truths cannot be accepted on the authority of experiences of human beings however exalted their status may be.

Some of the special features of Veda as a *pramāṇa* are further delineated in order to emphasize its separate status from other *pramāṇas* which operate in the realm of worldly realities. First, although revealed truth is extra-empirical (*aloukika*), its exposition is, nevertheless, always in terms of our empirical experience. The use of language for this purpose supports this conclusion. Secondly, any conclusion arrived at on the basis of this new testimony should not be capable of being contradicted by other *pramāṇas* such as perception and inference. Thirdly, it should be possible to identify some pointers to the truths proclaimed by the Vedas on the basis of reasoning which we normally employ. We have come across such pointers during the course of our earlier discussion; establishing the high probability of existence of the fourth state of consciousness through an analysis of the totality of our existence is one such instance.

4.6 Evolution

The *Vedāntic* thesis declares that pure consciousness is the common substratum for both spirit and matter. Veda gives precedence to spiritual evolution over biological evolution. However, objections to this thesis have been raised by rival schools of philosophy based on the premise that because consciousness is a sentient principle, it cannot possibly give rise to the insentiency that we observe in the universe. Many schools of philosophy that traditionally rival the *Vedāntic* thesis consider this disagreement as one of the chief obstacles for possible reconciliation. These schools subscribe to the *Upaniṣidic* truths but present different interpretations about the interrelationship between matter and spirit. In the classical thesis of *Vedānta*, there are several reasons advanced to demolish these spurious objections and assert that it is spirit that also gives rise to matter. We shall present some of the prevailing views on evolution of consciousness in order to provide some further insight into this subject.

According to modern scientific thinking, the claim is that life has arisen out of a primordial soup which is made up of various chemicals. In fact, the question of ' what is life' is understood to have been satisfactorily answered by the chemists. Biological evolution assumes that life has arisen out of insentient matter. However, science has not yet provided an unequivocal answer to the question of how consciousness and self-awareness appeared on this planet after the onset of life. In chapter 1, we speculated on the possibility that the new theory of chaos may suggest a way of answering the question because of its ability to handle problems involving organized complexity. In the case of consciousness, the premise is that it pertains to the manner in which molecules are arranged in a complex hierarchy rather than to the properties of the molecules themselves. If future research were to confirm this hypothesis, in-sentiency would still be the cause of sentiency, which is exactly the reverse of the *Vedāntic* thesis.

In this connection, it is interesting what George Wald, a Nobel Laureate in Medicine, has to say about a plausible solution to the question of a relationship between consciousness and evolution. In his discussion [42], the scope of the term consciousness is initially limited to the human mind. First of all, he refers to the monumental work of Wilder Penfield, the great Canadian brain surgeon who wrote *The Mystery of the Mind: A critical study of Consciousness and the Human Brain* [29]. What catches his attention is Penfield's conclusion that he could not, in his numerous experiments in brain surgery, locate the seat of consciousness in the human brain. As such Penfield considers the two fundamental elements, namely the mind and the brain, as two semi-independent elements. Wald goes one step further and concludes that " the mind is not not only locatable, *it has no location.*" He goes on to say, "What would it mean to assert that something exists for which we have no *evidence*? We encounter here the deep ambiguity between *being* and *being known.* Our consciousness is not alone the precondition for science, but for reality: what exists is what has become manifest to our consciousness." Wald proceeds to connect this riddle with an entirely different riddle posed by the anthropic principle which dwells on the peculiar character of the universe that breeds life. The connection that he speculates to be true makes very interesting reading:

> A few years ago it occurred to me that these seemingly very disparate problems might be brought together. That would be with the hypothesis that the mind, rather than being a very late development in the evolution of living things, restricted to organisms with the most complex nervous systems–all of which I had believed to be true–that mind instead has been there always, and that this universe is life-breeding because the pervasive presence of mind had guided it to be so.

Wald's conception of a non-locatable mind comes close to the *Vedāntic* concept of pure consciousness.

In addition to the views expressed on the cause and effect relationships between sentiency and in-sentiency, we also mention a third view which is also extant in some of the schools of philosophy. This view is that spirit and matter always coexist, an observation which is patently true on the basis of human experience. We know that the sentient mind and the insentient physical frame coexist, and we do not ask for further proof for validating this assertion. From this actual experience of being and becoming at the level of the individual mind and by means of the usual parallelism that we seek between the individual mind and the universal mind, we entertain the distinct possibility that spirit and matter also coexist at the level of the cosmos. This view, however, skirts the problem of directly addressing the causality issue between spirit and matter. In any case, acceptance of this view would mean that one cannot emphatically endorse the prevalent scientific view that consciousness is a phenomenon that appears only much after the appearance of life on this planet. This view about the coexistence of spirit and matter at the level of the cosmos is in conformity with George Wald's speculation on the subject which we have referred to earlier.

We conclude this discussion about spirit and matter by making the observation that *Vedānta*'s main focus is only on spiritual evolution and not on biological evolution. *Vedānta* emphasizes that it is the spiritual element that gains precedence in defining the interrelationship between spirit and matter.

4.7 Omega Point

Teilhard de Chardin, the late French Paleontologist and a Roman Catholic priest of the Jesuit order, came up with a thesis called *Omega Point* in an attempt to reconcile the Biblical view of creation with Darwin's theory of evolution. The latter theory was, and still is, considered to be contrary to the beliefs of Christian theology. As a punishment for having tread on a highly controversial ground, Chardin was sent away from France to China with strict instructions not to publish his heretical ideas. He eventually returned from China to New York where he lived till his death in 1955. It was only after his death that his friends and admirers published an English translation of his book, *The Phenomenon of Man*, with an Introduction by the British biologist Julian Huxley.

Chardin visualized the total energy of the universe comprising matter and spirit in two forms: first, the *tangential form* to represent the purely physical energies and second, the *radial form* to represent the spiritual energies within man. The recognition of the radial form of energy was very significant since the cosmological theories of evolution have invariably been based solely on a lifeless universe without any suggestion whatever of the possible impact of life on evolution. Evolution, according to Chardin, had to take into cognizance both the tangential and radial forms of energy which was a clear departure from the traditional view. Furthermore, he was more concerned with the future evolution of the universe rather than the evolution from the beginning of time to the present. *Chardin's concept of the radial energy was that it represented an evolutionary mechanism, that is, a non-physical entity for purposes of guiding evolution.* The British biologist Julian Huxley has written the following in his Introduction to Chardin's book: ' Teilhard, extrapolating from the past into the future, envisaged the process of human convergence as tending to a final state, which he called the Omega Point as opposed to the *Alpha* of elementary particles and their energies.'

Chardin's concept of the Omega Point has significance to both theology and science. The theological significance is that Omega is considered as the scientific equivalent of the God of Christian theology which is the point of contention from the Biblical view of creation. The scientific value of the Omega concept remained dormant until some renowned scientists like J.B.S. Haldane, J.D.Bernal, Paul Dirac and Freeman Dyson considered the hypothesis of the prevalence of life till the end of time as extremely significant to the future evolution of the universe. In this connection, the contribution of the Princeton Physicist Dyson, whose work we have extensively referred to in chapter 1, has to be singled out for its importance since he has worked out a detailed mathematical model of evolution based on the assumption of life till the end of time. Dyson also recognizes the contribution of the physicist Jamal Islam of Bangladesh to his thinking on the subject. The Dyson model can be regarded as the scientific version of the Omega Point theory.

We had earlier commented on the concept of ' heat death' arising as a consequence of the second law of thermodynamics which is based exclusively on the evolution of physical energy of the universe. Many scientists and philosophers have commented on the underlying pessimism of this view of the future of the universe. To quote Charles Darwin [18]

> ... the view now held by most physicists, namely that the sun with all the planets will in time grow too cold for life, unless indeed some great body dashes into the sun and thus gives it fresh life–believing as I do that man in the distant future will be a far more perfect creature than he now is, it is an intolerable thought that he and all the other sentient beings are doomed to complete annihilation after such long-continued slow progress.

In contrast to the dire prediction based on the concept of heat death, the philosophical underpinnings of scientific theories concerning Omega Point are based on a great deal of optimism for the humans since there is an assurance of the continuation of life. The Omega Point theory also implies that there will be sufficient technological progress to enable humans to extend life into the rest of the universe. This is an important corollary because the demise of our own planet is expected to occur much sooner than the 'ultimate future' corresponding to the Omega Point. The time corresponding to the Omega Point can be worked out in mathematical terms depending on the cosmological model used, but it is sufficient for our purposes to recognize Chardin's interpretation that the Omega has both an evolutive aspect and a transcendental aspect. The former is in time whereas the latter aspect is beyond time. What is in time is in the manifest field of existence and what is beyond time is in the non-manifest transcendental field. According to Chardin: ' Autonomy, actuality, irreversibility, and thus finally transcendence are the four attributes of Omega.'

The concept of Omega and its attributes are very close to some of the metaphysical concepts of *Vedānta.* Swami Ranganāthānanda [34] says:

These attributes of the Omega tally in essentials with the attributes given to *mahat* in *Vedānta.* As the highest reach of the value of personality, the *mahat* is known as *Hiraṇyagarbha* or cosmic Person, the 'Self-born', of whom only one quarter is (in time and) expressed in cosmic evolution, while three quarters are ever transcendent and immortal.'

We shall again refer to the concept of Omega in chapter 5 when we discuss the concept of *mahat* in Saṅkhya philosophy.

We shall conclude this section by making some comments on the physicist Frank Tipler's latest thesis on 'the physics of immortality' which also has its origins in the study of the ultimate future based on the *postulate of eternal life.* Tipler makes some sensational claims. To cite only a few: ' All concepts, omnipresent, omniscience, omnipotent, resurrection, Heaven etc., will be introduced as purely physical concepts.' ... 'no appeal, anywhere, to revelation.' ..... ' human soul is nothing but a specific program being run on a computing machine called the brain. It is considered as the software.' .... ' Eventual Return says that all events in nature repeat themselves in exact detail again and again.' .... 'Omega Point Theory provides a plausible mechanism for a universal resurrection.' .... 'If any reader has lost a loved one, or is afraid of death, modern physics says: Be comforted, you and they shall live again.' The Omega Point also implies immortality in the Biblical sense of resurrection of the dead to eternal life. Tipler makes the extraordinary claim that the study of metaphysics can be encompassed by the discipline of physics. He backs up his thesis with involved mathematical calculations that would need a detailed understanding of several scientific disciplines where only experts in those fields can vouch for the authenticity of his conclusions. Although his main objective is to validate the central assertions of Judaeo-Christian theology through the discipline of physics, the entire tone of his narration is bound to invite some healthy skepticism about his conclusions. It is next to impossible to establish correspondences with any of the Vedic concepts since there is no room in Tipler's thesis to incorporate ideas arising from mysticism.

4.8 Artificial Intelligence and Consciousness

The common meeting ground for physics and metaphysics is in the study of consciousness. While metaphysical inquiries into consciousness proceed from the firm conviction that the phenomenon is beyond the grasp of physics and other scientific disciplines in general, there are physicists, on the other hand, who have never abandoned the hope of understanding consciousness in purely scientific terms. This search for the missing science of consciousness' [31] is propelled by the physicist's realization that his view of the universe would be utterly incomplete without a satisfactory explanation of consciousness. There are two extreme viewpoints that are prevalent amongst physicists on the subject of consciousness. First, there are those who argue that it is only a matter of time before science can fully comprehend the study of consciousness, which, in effect, is also a claim that physics has the undoubted potential for encompassing metaphysics. This optimistic appraisal is based on the assumption that the promising study of the complexity of the human brain which is proceeding at a rapid pace will eventually provide the key for the understanding. Physicist Frank Tipler, in his book on *The Physics Of Immortality*, has already declared that physics has put metaphysics out of business! As opposed to this view, there are other scientists who argue that the study of consciousness will forever elude the grasp of science because of its exclusive reliance on quantitative methods and calculational procedures. This view is based on a careful observation of the full potentialities of the human mind including those of the unusually great contributions of human genius throughout recorded history which defy any rational explanation for their occurrences. These observations have led to the conclusion that there is more to consciousness than what mere scientific calculations can unravel.

Roger Penrose, an eminent British mathematician and cosmologist, has made an extensive study of the problem of consciousness in his book *Shadows Of The Mind* [31] which is a sequel to his earlier widely-read book *The Emperor's New Mind* [30]. His main objective is to cast serious doubts on the assertions of strong *artificial intelligence* (AI) which is a subject within computer sciences, and to suggest that there is yet a missing ingredient to be discovered before a physicist could complete his understanding of consciousness. The book sheds a great deal of light on the grey area between physics and metaphysics from the vantage point of a mathematical physicist. In order to appreciate the essence of Penrose's thesis, it is necessary to comment briefly on the basic ideas underlying AI.

It is now common knowledge that the digital computer is having a pervasive influence on all branches of study, and in all spheres of life. Computers have drastically changed the manner in which we acquire secular knowledge and conduct our worldly transactions. There is a vast array of electronic gadgets with computers as their main components which provide machine intelligence in order to considerably enhance the capabilities of man. The revolution in material sciences has also enabled the production of extremely powerful computers in unimaginably tiny sizes to assist the versatility of the applications. These *intelligent robots* have not only enhanced the capacity for man's sensory perceptions but also they have greatly expanded the range of the human mind. Routine tasks have become much easier to perform as a result of the combined man-machine intelligence. We have arrived at a stage in our civilization when it has become inconceivable to think of making policy decisions on complex problems without the aid of computers, the underlying assumption being that the vastly superior computational power of the machine is an inevitable accessory for making sound decisions. In other words, it is felt that man without the invisible aid of the 'thinking machine' would suffer a grave inadequacy for performing complex tasks.

But there are computer scientists who claim that computers can do much better than merely serve as aids to human intelligence. Their claim is, that given time, computers can grow out of their present subordinate roles because they have vested in them the immense potential for acquiring AI which would place them well above the level of ordinary human intelligence. Furthermore, these proponents of strong AI hold out the promise that both the gnawing problems of the world at large and those of the internal universe of man have a better chance of successful resolution when we have such powerful entities endowed with such vastly superior machine intelligence in our midst. There is also the explicit assumption that superior computational power of the latter entities can also simulate human consciousness.

There are several questions which Penrose raises on the subject of superior computation and consciousness underlying the claims of strong AI. Is intelligence solely dependent on computational power, or is there some other factor which is contributing to it? How can our innate feelings of pleasure and pain and such other duality of feelings fit into a theory based exclusively on computational power? Does it make any sense to envisage the possibility of computers having *minds*? Can science deal with questions relating to human consciousness? Penrose presents four different viewpoints on this controversial subject:

A. All thinking is computation; in particular, feelings of conscious awareness are evoked merely by carrying out of appropriate computations.

B. Awareness is a feature of the brain's physical action; and whereas any physical action can be simulated computationally, computational simulation cannot by itself evoke awareness.

C. Appropriate physical action of the brain evokes awareness, but this physical action cannot even be properly simulated computationally.

D. Awareness cannot be explained by physical, computational, or any other scientific terms.

The viewpoint expressed in **A** is the strong AI view. In effect, it is based on the understanding that there is a direct link between a physical process and computation . To quote Stephen Wolfram [17] of the Institute of Advanced Studies in Princeton: " Physical systems are viewed as computational systems, processing information much the way computers do." The physical system of interest in the discussion of **A** is the human brain because consciousness is presumed to be the result of its functioning.

To understand the relation between a physical system and its corresponding computational system, consider the motion of the Earth round the Sun. If we note the initial values at time t for the position and velocity of Earth revolving round the Sun, their values at a later time T can be determined by the motion of Earth subject to the mathematical laws pertaining to rigid body motion. Consequently, the physical process of the Earth itself can be viewed as a simulation much akin to a digital computer where the input-bit string of ones and zeros corresponding to the initial values of position and velocity at time t are converted into an output-bit string of ones and zeros corresponding to the values at time T. It is from this viewpoint that a physical entity is considered as a pattern of information, subject only to a set of mathematical laws, thus eradicating any distinction between a physical entity and a computational entity.

Based on the above argument, the strong AI proponents come to the astounding conclusion that a simulated mind would be conscious. Penrose casts serious doubts on the validity of the assertion of the strong AI view because, according to him, the physical process associated with consciousness is not exclusively computational in character but also involves a non-computational entity. However, he claims that this latter entity can be identified based purely on a knowledge of physics.

The viewpoint expressed in **D** is quite the opposite of the strong AI view expressed in **A**. It holds that the study of consciousness is beyond the reach of science in general. Penrose calls this the mystic view, a view that is shared by no less a mathematician than Gödel who is considered to be one of the greatest mathematicians of the last century.

We shall briefly comment on the above mystic view with reference to our study of Vedic philosophy. The reason that Vedas are considered the real testimony (*pramāṇa*) for the existence of pure consciousness (*ātman*) is because of the conclusion that it cannot be validated on the basis of testimonies that are invoked for validating scientific truths. In effect, this is an alternate statement of **D**. Our distinction between *parā vidyā* and *aparā vidyā* is also in the same spirit as the statement in **D**. *Ātman* is the spiritual element which coexists with all other states of consciousness of our ordinary existence. Specifically, there is no mention in the scriptural literature as to the location of this spiritual element in man despite its proclaimed immanence.

Penrose's rejection of the strong AI thesis is based on his conviction that human understanding cannot be an entirely a computational or an algorithmic activity. He bases his arguments from the vantage point of mathematics where the "thinking processes have the purest form." If the conclusion is shown to be true in mathematics then the generality of the result follows in any other form of thinking. Penrose draws attention to the full implications of Gödel's theorem in support of his stand. According to Penrose, this latter theorem is not only a very fundamental result in logic and mathematics, but it also sheds light on the philosophy of the mind. He says: " Thus, Gödel appears to have taken it as evident that the *physical* brain must itself behave computationally, but that the mind is something beyond the brain, so that the mind is not constrained to behave according to the computational laws that he believed must control the physical brain's behavior."

Penrose invokes the Gödel argument to summon concrete support for his stand that the strong AI view is flawed because it rests its case totally on computational power and algorithmic capability. However, he is not inclined to embrace the mystic viewpoint expressed in **D** which precludes the study of consciousness by physics. His thesis is that one has to include a non-computational and non-algorithmic element in the study of consciousness. Also, he strongly asserts that this missing element can one day be comprehended by the discipline of physics and proceeds to suggest a line of research which has the promise of completing the jig-saw puzzle.

Perhaps the most pertinent part of Penrose's book that is germane to our study of Vedic philosophy is the manner in which he invokes the Gödel argument to establish that " human insight lies beyond formal argument and beyond computable procedures" and in the explicit statement contained in **D**. The philosophical import of Gödel's theorem will lend a great deal of intellectual clarity to the metaphysical doctrines based on the Vedas and, in particular, when we discuss *advaita* in chapter 6.

4.9 Some Comments On Cognition And Creation

The *Vedāntic* concept of creation is from the vantage point of states of consciousness including the Self. It is quite distinct from what we normally mean by that term in cosmology. Cosmological theories, whether of modern or ancient vintage, assume historicity to the idea of creation, as is the case in any scientific study. In the *Vedāntic* thesis, what is meant by creation is its direct relation to the faculty of cognition. All that is necessary for the philosophical discussion is to recognize that creation and cognition are concurrent phenomena. In other words, there is perfect synchronism between the ' creation of the world' in the sense just referred to and its corresponding cognition. The world originates when there is cognition of it; when there is no cognition, as in deep sleep, there is total dissolution. The same conclusion is held true with respect to the origination and dissolution of any particular object.

We quote from Arthur Osborne's book on *The teachings of Bhagavān Śrī Ramaṇa Maharishi* [27] in this context. The great sage of Arunācala, which is close to the city of Tiruvannamalai in Tāmil Nādu, is considered by many as one of the greatest saints of twentieth century India. He taught the philosophy of non-dualism on the basis of his spiritual experiences which he had already undergone even prior to his coming into contact with the appropriate sacred texts of this school. The following remark is attributed to the Maharishi:

> *Vedānta* says that the cosmos springs into view simultaneously with him who sees it and there is no detailed process of creation. It is similar to a dream where he who experiences the dream arises simultaneously with the dream he experiences. However, some people cling so fast to objective knowledge that they are not satisfied when told this. They want to know how sudden creation can be possible and argue that an effect must be preceded by a cause. In fact they desire an explanation of the world they see about them. Therefore the scriptures try to satisfy their curiosity by such theories. This method of dealing with the subject is called the theory of gradual creation, but the true spiritual seeker can be satisfied with instantaneous creation.

Before making further comments, we recall that there are three states of consciousness which define the totality of our existence: the waking state, the dream state and the deep sleep state. Furthermore, the fourth state of consciousness (*turiya*) coexists with all the three states. That means that at any given instant we will be in one of three pairs of consciousness: Self and waking state; Self and dream state; and Self and deep- sleep state, although the Self is not experienced because of *avidyā*. The junction point of these three states is always the Self only because the union of any two of the pairs will result in the pair *Self– Self*. The perception of change in the states of waking and dream states is directly accountable to the constant prevalence of a stationary consciousness represented by the Self. It is the analysis of cognition in these states with their corresponding origination that are important to philosophical inquiry.

4.10 Space, Time and Causality

First, we shall comment on the law of causality as it operates in the realm of worldly realities in order to clearly distinguish its meaning from its operation in the philosophical realm. It is common knowledge that for an intellectual understanding of an empirical phenomenon, we always try to establish the cause that is producing the effect. The cause that we look for is local in character in the sense that we investigate for a cause that is immediately antecedent to the effect. It is the cause and effect relationships that are central to the study of scientific determinism which we discussed in chapter 1. The majority of problems appearing in science and technology are based on deterministic theories. Furthermore, cause and effect relationships are of interest in almost all disciplines including, of course, in our ordinary worldly transactions. Normally, we distinguish between two types of causes: the material cause and the effective cause. For example, a nice piece of sculpture is the effect for which the material cause is the type of stone from which it is carved, and the sculptor himself is the effective cause.

But the principle of law of causation (*kāryakāraṇabhāva*) has an entirely different meaning in philosophy. While discerning worldly realities, when faced with the problem of finding the cause B for effect A, we are not normally interested in pursuing the line of inquiry further to find out the cause C for producing effect B. For most types of analyses of practical problems, we are satisfied with the immediately preceding cause. In our earlier example of a piece of sculpture, we would not be interested, for instance, in asking ourselves what the material cause was for the type of stone used, and continue in that vein on and on, *ad infinitum*. In philosophical reasoning, on the other hand, we are not merely satisfied with determining the antecedent cause, but want to probe in depth to discover the true substratum of the effect under investigation. The algorithm for arriving at the ultimate basis of an effect is based on some key subtle observations. First, we do not distinguish between a material cause and its effect because the latter exists only when the former exists.

This is the criterion of non-difference between cause and effect. Undoubtedly, the effect and its cause will invariably have different names and forms. The piece of sculpture certainly has a different name and form from that of the stone it is carved out. Taking this observation into account, we only deny the difference between cause and effect without making the ludicrous assertion that the two are identical in name and form. This suggests the second criterion of not claiming identity between cause and effect.

We further notice that what is common to all the series of effects is their existence (*sattā*) at each stage in the long chain of causes and effects. That every cause and every effect exists is an undeniable fact. Existence is therefore the underlying reality of any effect, a philosophical discovery which is of interest to us. And then we come to the grand conclusion that *Brahman*, which is both the First cause and the First effect, whose intrinsic quality is its primal existence, is the material and effective cause for all the effects that we observe in the universe. Furthermore, we are assured that *Brahman* cannot be the effect of some other material cause because it can never be the object of investigation. Such a false inquiry would only result in an infinite regress. *Brahman*, as we know from the scriptures, is one without a second. A cause and effect relationship can be entertained only when there is a feature that can clearly distinguish between the two, and there is no such distinguishing feature in the case of *Brahman*.

The *Vedāntic satkāryavāda* is the philosophical principle of causation which asserts that the cause, in fact, is in the effect itself, based on the discovery that there is no distinguishing feature between them because what is common to both is existence. All the effects in the sequence can all be viewed as modes of *Brahman* which is the primal existence. From this law of causation, one can also conclude that the notion of becoming which represents the transient character arising from the notion of change is, in the ultimate analysis, also fictitious in character. To illustrate the property of becoming, one can use the analogy of a movie which is the result of the quick motion of a series of stationary snapshots taken at successive instants of time. At each instant of time over the duration of the transient phenomenon, what we have at the substratum, in fact, is being, which is existence itself.

We now take up the discussion of time leading up to some philosophical ideas pertaining to this vital concept. We have, in chapter 1, referred to the idea of a 'flow of time' underlying the notion of becoming and connecting our real experiences about events of past, present and future. This is what defines the psychological arrow of time. There is also the notion of irreversibility as evidenced in the discipline of thermodynamics where time enters in such a way that we cannot restore order once it is disturbed. Furthermore, the disorder in this universe is steadily on the ascendant and will continue until such time, billions of years ahead of us, that the entire universe suffers from dissolution, which is what is technically termed as 'heat death'. This dire prediction also defines another arrow of time as a result of the steady expansion of the universe. Time assumes dimensions from light years as appropriate for the study of our cosmos to very minute dimensions as required in the study of particle physics. Time and its allied concept of space are quite central to mathematical models which describe various theories of worldly phenomena. We need the concept of time whether to understand dynamic motion of a space shuttle or to gauge the speed of operation of a digital computer or the speed of light. Apart from such esoteric references to time in several applied areas, we know that the concept of time is indispensable to the conduct of our daily transactions. Budgeting time has come to be considered as a mark of efficiency in our various endeavors.

Turning our attention to the philosophical implications of time, the *Vedāntic* view also confirms the scientific view that the concept of time has absolutely no meaning before the 'event of creation' of our universe. It is not even defined in that context. It acquires meaning only after that event if that magical moment has any meaning at all strictly on a time scale; some well-known physicists speculate that it is a space-time event which is a concept described by Einstein's theory of relativity. That is why the meaning of eternity is not from everlasting to everlasting, but rather something that transcends the concept of time altogether. In *Vedānta*, time is considered as a pseudo relation that exists between the transcendental realm and the realm of empirical reality. That is why the *Vedāntists* conclude that while time has no beginning at all, it nevertheless, has a definite end because it has its origin in eternity and an end at the moment of termination of *avidyā*. The discussion so far has dealt with the role of time in our external universe for which we can, at best, have only intellectual understanding and not actual experience. For this, we have to turn to the manner in which time manifests itself at the human level.

Time is experienced in the waking state and also in the dream state, albeit in a distorted way. We say it is distorted since we consider the experience of space, time and law of causality in the waking state as the true state of reference for judging all worldly reality. However, time is not experienced at all neither in the deep sleep state nor in the state of *samādhi*, which is the higher state of consciousness. Also, the universal experience ' I am present' in the absence of either sensory data or any discernible cause for the experience suggests that that knowing is experienced directly by the *ātman*, which is the eternal witness. In fact, careful analysis of the experience of time in the totality of our existence will provide yet another trace of the truth about the transcendental realm.

As for space, the important observation that has to be made is that although it is a phenomenon closely allied to time, there is however, an important difference between the two. Space is something that evolves after the 'moment of creation', and it is therefore included in time. Consequently, there is a clear separation between the realm of *Brahman* which is marked by the absence of time and the empirical idea of space which has meaning only when consideration of time enters into the picture. This *Vedāntic* notion is also in accord with the findings of modern-day physics. We have a further philosophical concept of *ākāśa* associated with space. The *Vedāntic* view is that *ākāśa*, which is loosely translated as 'ether', is contained within space and has sound as its characteristic feature. What we see all around us in the open skies is *ākāśa* and not space. In *Vedāntic* terminology, *ākāśa* is considered a limiting adjunct (*upādhi*) on space, and that limitation is what serves the empirical need of defining directions such as north, east etc. The concepts of space and time are not established on the basis of any *pramāṇa*, and as such, the scriptures also do not make any specific mention of them.

4.11 Traces Left Behind By The Supreme Truth (Brahmavāsana)

There are indications in our worldly experiences which, if carefully analyzed, would establish the plausibility of existence of the unseen transcendental reality. Such pointers serve a very useful purpose in the exploration of the unknown. We shall list some of these pointers which have appeared at various places in our earlier discussion for the sake of ready reference.

- Perhaps the best indication is the analysis of the totality of our existence in the three states of consciousness. When the question 'who am I' is pursued in the waking, dreaming and sleeping states, we come to the conclusion that there is an eternal witness, the fourth state of consciousness (*turiya*) which coexists with each one of the three mutually exclusive states of consciousness. The *Māndukya Upaniṣad* gives an excellent analysis of this indication.

- In our section on body-mind relationships, we pointed out that there is a significant amount of research that establishes correlation with the state of consciousness prevailing under conditions of meditation. There is conclusive evidence to prove that the metabolic indications of this state of consciousness are very different from those for the ordinary states of consciousness. This evidence provides another pointer indicating the possibility of the existence of a fourth state of consciousness independent of the other three. Furthermore, the investigations reveal that this state coexists with one of the other three states of consciousness at any given time.

- There is an universal human urge to be at all places at the same time, to know everything, and to always be happy. These three urges which are summarized as ' to be' ' to know', and ' to be happy ', are considered to be dim reflections of the essential nature of *Brahman* which are existence, consciousness and bliss (*saccidānanda*). These basic instincts of man are also responsible for producing an innate fear of death, fear of ignorance and fear of misery.

- *Jīva* (soul) is invested with the faculty of *jñāna* (knower) and *kriyā* (doer) along with some limitations. Again, these are qualities of the personal God *Īśvara* who has the infinite power of knowing and acting without any constraints imposed on Him.

- All objects of our experience are characterized by their existence (*sattā*) and shine (*prakāśa*). These are true even in the dream state. An analysis of these features will lead us to the conclusion that it is the effulgence of the Self that is responsible for these manifestations.

- We have earlier discussed the psychological arrow of time in terms of past, present and future events. The notion of time as experienced by man is different from the physicist's conception of it. When we say present, it usually refers to an interval of time. In this regard, we find that the notion ' I am present ' is universal. This notion does not even depend on a cause for producing that feeling. This feeling of 'now' of our worldly experience is considered to be a pointer to the ' eternal now ' of the transcendental reality. The philosophical concept of time as a pseudo-phenomenon connecting the realm of eternity where time is not even defined and the universe of our experiences where time is very much a factor leads us to the conclusion that time itself could be a definite pointer towards the ultimate truth.

4.12 Aesthetics and Spirituality

Aesthetic pleasure is derived from the beautiful in both nature and art. It has a bearing on spirituality since true aesthetic experience is known to give a foretaste of spiritual experience. Unlike Western philosophy, Indian philosophy does not include aesthetics as one of its sub-disciplines, however, the latter accords the same importance to beauty as it does to goodness. The reason is simply that the final goal of philosophy in its practical aspects is liberation (*moksha*), which is outside the scope of aesthetics. Indian philosophy places both beauty and goodness on the same pedestal and thereby suggests a strong connection between aesthetics and ethics. Aesthetics is given the status of an independent discipline that comes under the purview of a group of professionals called the *Alamkarikas* (linguistic critics), who freely pursue their own ideas. This discipline is primarily focussed on poetry and drama, but its central findings are applicable to all forms of art. The Beauty in nature does not occupy these critics since it receives adequate treatment in philosophy, either explicitly or implicitly. The final aim of this discipline as propounded by the *Alamkarikas* is to investigate the conditions under which one could attain a state of unalloyed pleasure from the experience of art. This latter kind of exalted tranquillity of mind shares some features in common with spiritual experience but is distinct from it in many important respects. Suffice it to say at this stage that a particular response to art is not considered a prerequisite for a spiritual aspirant to advance towards *moksha*.

Considering aesthetics with its sole focus on art experience has some decided advantages. The freedom enjoyed by the new discipline guarantees that it is not subject to the constraints imposed by a specific philosophical inquiry. For instance, the discussion of the experience of art does not have to minister to the doctrines of Vedic philosophy with its central concepts of *ātman* and *brahman*. The same holds true for the numerous branches of philosophy of both the East and the West. Epistemological considerations, while they constitute an essential part of philosophy, they are, however, peripheral to the theory underlying the experience of art. Instead, the main concern of aesthetics is merely to examine conditions under which one could derive a blissful experience, albeit for short duration of time, in the strife-ridden world that we are living in.

The distinctive nature of art can be highlighted by considering some of the principal differences between beauty in nature and beauty in art. The Vedic view of nature is that it is the very personification of beauty when it is perceived in its entirety such as when we consider Personal God as the very embodiment of nature. Fragments of nature, on the other hand, reveal only beauty combined with some ugly features. A holistic experience of nature is a transcendental one, which is privy only to self-realized souls, an extreme rarity amongst human beings, who have attained spiritual perfection by pursuing a rigorous spiritual discipline combining a steady contemplation of the ultimate reality with an arduous cleansing of the doors of perception. For most others, who constitute the bulk of humanity, the human condition is such that nature can only offer a mixed experience of both the beautiful and ugly because of the fact that human perception is fogged by varying degrees of spiritual ignorance. Ordinary mortals can only look at the world with their own special rose-colored glasses, which denies them the ability to perceive perfectly. Despite this serious handicap, there is a remarkably optimistic finding: even under realistic and imperfect conditions of perception, the artistic medium can provide an escape hatch from the dual experience of pleasure and pain and thus elevate one into the higher realm, where only pure pleasure is experienced. Furthermore, there is the unambiguous assertion that no medium other than art possesses this unique capability of uplifting one into a virtual universe, which is void of worldly displeasure.

The main characteristics of the ideal experience of art are: a) a disinterested contemplation of beauty, i.e., a complete suspension of egocentricity and b) the experience of a kind of pleasure that is reminiscent of the spiritual experience articulated by self-realized persons. Both disinterested contemplation and attainment of pure pleasure have striking similarities to spiritual experience.

However, there are also some essential differences between the two types of experience which make them quite dissimilar. Aesthetic experience, unlike spiritual experience, is induced by an external stimulus, be it poetry, drama, music, painting, or one of its myriad forms of art. Once the stimulus is withdrawn, aesthetic experience also suddenly comes to an abrupt end; the virtual universe of the artist's creation in which the recipient is a full participant dissolves like the kingdom of Cinderella at the stroke of twelve. Pleasure through art is ephemeral, unlike spiritual experience, whose hallmark is permanence; the latter is not induced by an external stimulus, rather it is achieved through an inner transformation of the mind, resulting from a steady practice that combines intellectual discipline in contemplating the ultimate reality with practical discipline dedicated to leading a moral life.

It is in the discussion of the content of art that Indians have made a crucial contribution through which it is entitled to call itself a discipline in its own right. We shall first briefly outline some of the fundamental ideas pertaining to art about which there is a wide concurrence of views. It is generally known that any specific art form has its unique set of excellences that can be broadly categorized under form and content. For instance, the form of poetry is its musical language whereas its figurative ideas and sentiments constitute its content.

As a general rule, the form should subserve the content if it is to preserve the quality of the art. Since the content plays the primary role according to Indian aesthetics, it would be worthwhile to note some of its broad features. First and foremost, the content should draw from real-life considerations accompanied by a generalization thereof. The expectation is that the recipient echoes the feelings portrayed in the artist's work when such a generalization takes place. It would be unrealistic to expect such a communion between the artist and the recipient, if the content were merely to arise from the intellect of the artist rather than from his emotions. Along with the ability to extract the essence of the real-life situation by means of generalization, there is also a process of abstraction that takes place, a process that renders the experience free of the specific context, which is precisely what induces a mood of disinterested contemplation in the recipient. The process of abstraction is such that the integrity of the real-life situation is preserved by the artist even while depicting it as a fiction. When the two processes of generalization and abstraction are properly utilized by the artist, the spectator will regard aesthetic experience as a special form of play and not as work associated with intellectual analysis.

With the above background on the intricacies of aesthetic experience, we shall now examine the significant change that has occurred through Indian thinking. The *Alamkarikas* began to single out emotion as the true content of art and to regard everything else as only its outer vesture. The characters in a drama, for instance, become secondary in importance to the emotion that characterizes them. It is important to note that the said emotion refers to the emotion contained in the situation depicted by the artist and, specifically, is not the emotional response of the spectator arising from his imagination of what the artist is depicting. This type of emotion, which is intrinsic to the depicted situation is called *rasa.* In a drama, for instance, the characters and the story, important as they are, they, however, subserve the primary emotion that is depicted.

By carefully observing the behavior of human beings, the *Alamkarikas* have identified nine primary sentiments, or *rasas*: These are: *sringara* (love); *haasya* (humor); *karuna* (pathos); *krodha* (violence); *bibhatsa* (heroism); *virya* (firmness and steadfastness); *bhayankara* (fearfulness); *adhbuta* (wonder); and *shanta* (tranquillity). The final sentiment in this list was a late addition and was included after much discussion within the profession. The sentiments are primary in the sense that they do not depend on anything else for their experience, unlike the secondary feelings such as doubt or despondency which rely on extraneous explanations. Put another way, the experience of *rasas* is untainted by the faculty of intellect, whereas the same thing cannot be said of the secondary feelings. This special property of the *rasas* makes it possible for the artist to depict a situation pregnant with a primary emotion that can serve as a channel of communication with a cultivated recipient, called a *rasika*. Such a recipient possesses a corresponding emotion in his psychological makeup. A *rasika* is one who has an innate taste for aesthetic experience. Since the primary emotion cannot be directly communicated to the spectator, the artist portrays some select features in his creation to achieve an indirect means of communication, called *dhvani*. For instance, an expression on an actor's face can depict fear or love, as the case may be. The theory of *rasa* and *dhvani* is quite elaborate and, as stated earlier, is well-developed in poetry and drama with the understanding that there are universal features which are applicable to all forms of art. We have given an outline here to point out the place of aesthetic experience in the scheme of spirituality.

Aesthetic experience can be described as serene (*vishranti*) since the mind has attained the seemingly paradoxical state of restful alertness. All the restlessness of real- life experience temporarily vanishes during that period. It is this perfectly tranquil state of mind that suggests a connection between aesthetic experience and spirituality as understood in the Vedic philosophy.

We recall from our earlier discussion of Vedic philosophy two essential ideas for establishing a connection with aesthetics. First, the *ātman* doctrine brings into focus the facet of ultimate reality at an experiential level; thus, it removes all possible vagueness in the dual concept of the *brahman* doctrine pertaining to the external universe which is purely at an intellectual level. Second, Vedic philosophy proclaims on the basis of its concept of *jivanmukti* that it is possible to attain self-realization in one's own life-time without waiting for the hereafter, thus injecting an element of robust optimism in the spiritual process. One of the essential experiences of a self-realized soul is that he or she will always be in a state of unalloyed happiness (*ānanda*), which is an innate characteristic of the ultimate reality along with the two others, namely, existence (*sat*) and consciousness (*chit*). We can now strike a parallel between spiritual experience as enshrined in the Vedic philosophy and aesthetic experience by making the observation that *rasa* is the aesthetic equivalent of *ātman*. The vital difference between the two types of experiences is, however, that while aesthetic experience, derived from the enjoyment of *rasa*, and resulting in pure pleasure is ephemeral, the *ātman* is not contingent upon sensory stimuli and is therefore permanent.

It is possible to establish the final connection between aesthetics and other metaphysical constructs of Indian philosophy on the basis of the *rasa* theory. What is important to note is that aesthetics deals with emotional experience, so the connection sought with philosophy is at the level of spiritual experience and not with its epistemological considerations. Since all metaphysical constructs that come under the Vedic fold emphasize the values in life, it becomes possible to make the connection between aesthetics and spirituality.

Finally, we note the significant difference between what science can offer to the understanding of metaphysics as compared to what art can offer. In the case of science, we only make use of scientific paradigms as sharp pointers to the understanding of philosophical concepts, all within the domain of knowledge, though we deal with two completely different realms of knowledge. As we have already pointed out, the empirical world of science and the realm of transcendental knowledge of the ultimate reality are distinct. We have earlier introduced the concepts of *aparā vidya* (lower knowledge) and *parā vidya* (spiritual knowledge) to explain the jurisdictions of science and philosophy, respectively. In the case of art, the connection with philosophy comes at the level of experience. We establish the parallel between pleasure gained by aesthetic experience with the happiness gained by spiritual experience. Art is a virtual world where only pleasure is present, so it is different from the experience of the natural world, where there is an experience of both pain and pleasure and from the spiritual experience where both pleasure and pain are transcended.

Part II

Metaphysical Theories of Vedic Philosophy

Chapter 5

Sāṅkhya and Yoga

5.1 Sāṅkhya

In the last chapter, we discussed several basic concepts which are useful for understanding the Hindu perspective on science and spirituality. Here, we turn our attention to an orderly presentation of the Sāṅkhya and Yoga philosophies which come under the Vedic fold. We have several purposes in mind for taking up the discussion of the Sāṅkhya philosophical school first. First, Sāṅkhya is considered to be one of the oldest philosophies of India constituting a theory of unrivaled elegance dealing with matter and spirit. It has had profound influence on other philosophical schools of both the orthodox and heterodox categories, that is, those that come within the Vedic fold and those that are outside of it. The specialty of Sāṅkhya is that it does not invoke the concept of *māyā* in dealing with the interrelationship between spirit and matter. Nor does God enter into the thesis in an explicit manner, which in fact, provides room for the serious criticism leveled against it by the rival philosophical doctrines. But for the moment we shall withhold our comments on the legitimacy of such criticisms.

The exposition of Sāṅkhya will also enable us to bring to the fore several other philosophical concepts in a connected way. These will be extremely useful for the discussion of other schools of philosophy. Sāṅkhya's primary emphasis is on the realization of the ultimate truth through knowledge which is arrived at on the basis of philosophical reasoning. We place emphasis on philosophical reasoning in this context because of the inadequacy of scientific rationality for matters concerning ultimate truth. Secondly, the Yoga philosophy, which is a distinct philosophy on its own, serves a complementary role to Sāṅkhya because it deals with the practical aspects of spiritual practice. The Yoga philosophy has a pervasive influence on all other philosophies. While all other schools of philosophy depend on Yoga philosophy, the converse, however, is not true.

The name of sage Kapila is associated with Sāṅkhya, and its idealistic realism is rooted in *satkāryavāda* which is the philosophical law of causality that we have already discussed in the previous chapter. Although we do not have accurate historical records, it is inferred beyond doubt that his philosophy was known to Mahāvīra who is the principal prophet of the early Indian religion of Jainism and later, to Buddha, the founder of Buddhism. Some protagonists of Sāṅkhya would go to the extent of asserting on the basis of relevant texts that Mahāvīra and Buddha were influenced by their doctrine, Buddha more so than Mahāvīra.

Sāṅkhya assigns ultimate reality to both matter and spirit and accord them equal status. We said in chapter 2 in our discussion of Vedic philosophy that the cosmic principle, whose manifestation is the universe, is identical to the psychic principle which is the essence of the teaching of non-dualism. In this philosophy, however, we proceed from the dual concepts pertaining to spirit and matter. *Puruṣa* is the name given to spirit which is nothing but the ultimate indestructible awareness. Specifically, it does not mean 'man' as it means in ordinary usage. The doctrine admits the existence of a plurality of selves and accordingly, takes into account as many *Puruṣas* as selves. *Puruṣa* therefore refers to both the entity at the cosmic level as well as its manifestation at the individual level. As for matter, *Prakṛti* is considered to be the source of the physical universe. It is from this undifferentiated, non-manifest *Prakṛti*, that all the diversity of the universe unfolds. *Puruṣa* and *Prakṛti* are therefore the two ultimate realities of the Sāṅkhya philosophy. The coexistence of the dual principles of *prakṛti* and *puruṣa* is displayed in several phenomena such as the body (*śarīra*) and the vital life that sustains the body (*śarīri*), inertness (*jada*) and consciousness (*cetanā*) etc.

One of the basic doctrines of Sāṅkhya deals with the theory of causality. All the physical entities of the universe are considered to be effects due to differentiation of the undifferentiated *Prakṛti*. What is latent becomes manifest whenever an effect is produced by a cause. This is called *satkāryavāda*, the essence of which is that the effect is always identical with its material cause. While any distinct entity which has name and form can clearly be perceived, the primal source, *Prakṛti*, remains nonetheless forever inaccessible to perception. Since all the diversity of the universe emanates from *Prakṛti*, the primordial nature, the theory of causality is also sometimes referred to as *prakṛti pariṇāmavāda*. But the protagonists of Sāṅkhya insist that the correct causation is *satkāryavāda*. *Prakṛti* is considered to be made of three distinct, natural constituents called *guṇas*, which mutually interact with one another thus ensuring their inseparability. Consequently, at every stage, *Prakṛti* will be influenced by all three *guṇas*. Both the primal substance and its constituents are considered to be without beginning.

The *guṇas* are *sattva*, meaning purity; *rajas*, meaning activity; and *tamas*, which stands for inertia. Each one of these *guṇas* represents a physical reality, and obviously, the three realities have distinct, and indeed antagonistic characteristics. However, their interaction is such that they can produce a single harmonious outcome. The analogy that is usually given to stress this point is the example of a lamp flame which has for its constituent factors a wick, oil and fire. Although they have very different characteristics, they work in unison to produce a very desirable result. Furthermore, everything that is transformed from *Prakṛti* is also made of the three *guṇas*.

The whole process of evolution is explained on the basis of the transformation that takes place in *Prakṛti* due to the interplay of its three natural attributes in different proportions. When *Prakṛti* is in an undifferentiated form, the *guṇas* are considered to be in complete equilibrium. Diversification, which is the same as the process of manifestation of the physical universe, is due to the interaction of the three *guṇas* in different proportions. However, no *guṇa* can be entirely absent; only it can approach an infinitesimally small value. When the first *guṇa* of *sattva* is predominant, and the other two have very small values, the resulting combination makes a pure substance. At the opposite end of the spectrum, if the substance has *tamoguṇa*, the attribute of inertness, as the predominant factor, the substance will be impure. Since such combinations could be infinite, there is no limit to diversification. Evolution, in this sense, means a change of forms resulting from different combinations of the *guṇas*. We have already drawn attention to the fact that a static universe was never envisaged in Hindu thought, as is evident from the concept that evolution was always considered cyclical in nature. Evolution, *sṛṣṭi*, is always followed by dissolution, *pralaya*, in this scheme of things. When the stage of dissolution is reached, all the diversity of the universe is considered to become latent again in *Prakṛti*. Furthermore, it is envisaged that the cycle of evolution and dissolution proceeds in an endless rhythm.

The discussion of the twenty-three evolutes of *Prakṛti*, which we shall now enumerate, and the order in which they appear makes for one of the most fascinating accounts of Sāṅkhya philosophy. At first, one wonders why some of the cosmic evolutes of *Prakṛti* should have connotations in terms of the several facets of the human mind. Such a treatment will only make sense if there is some parallel that we can establish for the human mind with the mind of the universe. It will be useful to recall our discussion of a similar subject earlier on while dealing with the scientific view of philosophical problems. There, we considered the mind operating at three different levels: first, at the level of matter in particle physics, where it also behaves like an active agent; secondly, at the human level; and lastly, at the level of the universe. Although the existence of a universal mind was only a scientific speculation in that discussion, the possibility of it was discussed. Suffice to say for our present discussion, that the Sāṅkhya scheme does indeed assume the existence of a parallel between the individual mind and the universal mind.

Before discussing the evolutes, we should note that the three *guṇas*, which are constituents of *Prakṛti*, are also identified by attributes of the human personality. As we proceed with the discussion of the evolutes, it becomes clear that what is established is some sort of isomorphism, some one-to-one correspondence between the physical aspects of the cosmos and the physical aspects of the human being. The delineation of the evolutes proceeds from the implicit assumption that such a symmetry exists. Since, according to this treatise, the physical ingredients of a human being are also faithful traces of the primordial *Prakṛti*, the cosmic evolutes can legitimately be assigned names much closer to our understanding. In other words, the human being can be considered to be a microcosmic model of nature for the purpose of delineating the evolutes of *Prakṛti*.

The very first element which manifests itself from *Prakṛti* is intellect (*buddhi*), or (*mahat*), which is the subtlest element in the scheme. It stands at the juxtaposition of the non-manifest and the manifest realms of existence. In modern terminology, we can call it the control variable governing the functions of the rest of the evolutes which sprout from it in a hierarchical manner. The next element which evolves from *buddhi* is *ahaṁkāra*, which is the principle of individuation. It is the 'I' notion, the egoism, which explains the feeling of one's separateness from the rest of the external world. Without this evolute, there would be no notion of worldly reality and, consequently, no possibility for any philosophical speculation. The choice, the number and the order of evolutes is explained on the basis of the coexistence of matter and spirit, as in a human being. It is therefore reasonable to suppose that matter works in unison with the spirit by providing the latter with a complete set of aids to life's experience in their order of importance. For instance, without the aid of *ahaṁkāra*, there is absolutely no feeling of duality that is essential for the experience of worldly existence. Therefore its ranking amongst evolutes is quite high, next only to *buddhi*, which is the first evolute that is subtler than the subtlest. Next to *ahaṁkāra* is *manas*, which is an important mode of the mind, which is responsible for weighing pros and cons of a proposition. The three evolutes *buddhi*, *ahaṁkāra* and *manas* are together referred to as internal senses (*antaḥkaraṇa*) in order to distinguish them from the external senses of the sensory organs. The five sensory organs and five motor organs are the next set of evolutes. This completes a list of thirteen evolutes which descend from *Prakṛti*. In the Sāṅkhya schematic of evolutes of *Prakṛti*, there is a fork emanating from the evolute of *ahaṁkāra*. The evolutes of *manas* and the ten sensory and motor organs branch off from one side of the fork from the evolute of *ahaṁkāra*. A second branch of evolutes consists of five subtle and five gross elements of the objective world. The sensory organs sprout from the *sattva* aspect of egoism whereas the elements of the physical world emanate from the *tamas* aspect of egoism. We have already explained that this difference makes an important distinction, a difference between subtlety and grossness, between purity and impurity.

The evolution of the physical elements is in two phases. In the first phase, they remain as the basic constituents, the *tanmātras* or the subtle elements, and in the second phase, these subtle elements combine successively to form the gross elements. The simple elements are elemental sound (*śabda*), elemental color (*rūpa*), elemental touch (*sparśa*), elemental taste (*ruci*), and elemental odor (*vāsana*). There is no distinction made between substance and quality, unlike that made in some other philosophical schools. The gross elements are formed successively as follows: elemental sound gives rise to the inner essence of space, *ākāśa*; *ākāśa* together with touch gives rise to air, *vāyu*; *vāyu* combined with color gives rise to fire, *agni*; fire combined with taste gives rise to water, *pāni*; and finally, *pāni* along with odor gives rise to earth, *pṛthvī*. Space is identified by sound, air by sound and touch, and so on so that earth is identified by all the subtle qualities.

The total number of entities in the Sāṅkhya schematic, proceeding from the subtle to the gross, are twenty-five in number. They are *Puruṣa*, *Prakṛti*, and its twenty three evolutes. The following schematic displays the pattern of this primary transformation.

The secondary transformation relates to the manner in which all entities of the physical universe are formed from the primary ones. Some might arise from the branch where the source *guṇa* is *sattva* and some others from the second branch where the source *guṇa* is *tamas*. This difference will provide a critical distinguishing criterion between the entities. Specifically, there is no new principle involved in this secondary transformation. It is important to emphasize that the internal senses emanate from the *sattva guṇa* of *Prakṛti*, whereas the elements of the objective world arise from its *tamo guṇa*. Most of the ingredients of our own physical bodies

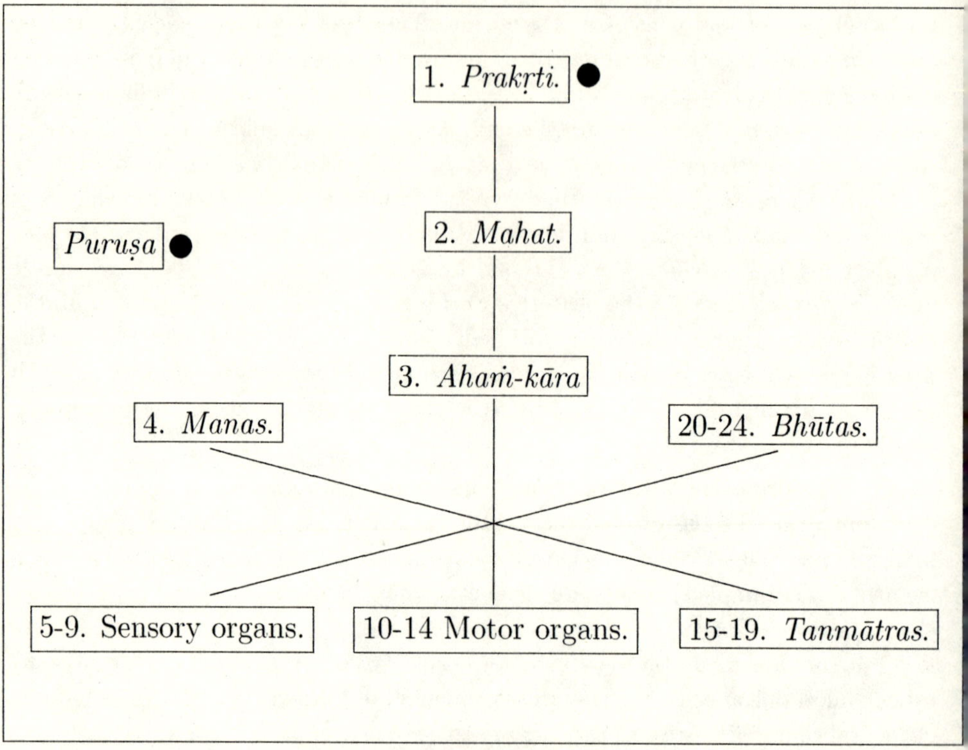

Figure 5.1: Sāṅkhya Schematic

belong to the latter category. It is interesting to note that even the three evolutes pertaining to the internal senses, namely, *buddhi*, *ahaṁkāra*, and *manas*, are physical in nature since they are all results of primary transformations of *Prakṛti*. But they are indispensable for the revelation of objects to the spirit, which is the same as saying that they are absolutely essential for mental life.

In our discussion in chapter 1 about the evolution of life and consciousness, we said that while the chemistry of life is well understood, the mechanism for the evolution of consciousness is not. However, we speculated that the new theory of chaos may hold a clue for the research on consciousness based on the premise that it has to do with the hierarchical manner in which the molecules of life are arranged, rather than the properties of the molecules themselves. The Sāṅkhya doctrine also assumes that the several modes of the mind are made out of physical ingredients but with the attribute of *sattva* predominant in their evolution from the primal substance of *Prakṛti*. This qualitative idea, however, does not provide any further insight into the vexing question about the formation of consciousness and self-awareness. However, it is interesting to note how the ancient doctrine of Sāṅkhya clearly distinguishes the manner in which the subtle and gross evolutes sprouted from *Prakṛti*.

The internal senses serve a dual role. First, they are fundamental for perception of external objects of the universe when they work in consort with the external senses, and secondly, they are also precisely the senses that have to be cleansed for achieving spiritual ascent. Moral and ethical cleansing, which are prescribed as prerequisites for the realization of the spirit, or *Puruṣa*, can now be understood as the cleansing of the internal senses whose composite characteristic is unique to an individual, as a result of different proportions of the three *guṇas*.

At this stage, we refer to one of the telling points made by sage Patanjali with regard to spiritual ignorance. He says that such ignorance is not a natural property of either *Puruṣa* or *Prakṛti* and therefore, it can successfully be eliminated. This change is possible because only a natural property of an object is indestructible, whereas a blemish resulting from a combination of two dissimilar elements can most certainly be destroyed. Spiritual ignorance belongs to this latter category. It is a blemish because it is a result of a combination of the two dissimilar elements of *Puruṣa* and *Prakṛti*. Ego is the result of a combination of the seer and the seen.

Returning to our main discussion, we note that the internal senses are inextricably linked to the *Puruṣa*, so that even the death of the physical body will not affect them. One can link this idea to the *Karma* doctrine which depends on the theory of transmigration. Although death puts an end to the mortal coil, the soul does not meet with the same fate. By virtue of this independent existence, the internal senses are called the subtle body, *linga-śarīra*, in order to distinguish it from the gross body, *sthūla śarīra*.

Earlier we said that the transformation of the ultimate reality of matter, *Prakṛti*, is not arrived at by any scientific investigation of nature, as in science. Rather, it proceeds from the understanding that the human being will treat himself as a laboratory for this investigation because of the coexistence of spirit and matter within him. We have already pointed out that the evolutes have cosmic as well as individual significance. This is particularly evident when we focus attention on the thirteen evolutes branching off from the *sattva guṇa* aspect of *Prakṛti*. In fact, in an individual life, the three cosmic *guṇas* of *sattva*, *rajas*, and *tamas*, take on the meaning of pleasure (*sukha*), pain (*duḥkha*), and bewilderment (*moha*). On the basis of this mapping of *guṇas* from the cosmic to the individual, one can conclude that an individual's aspiration should be to make *sattva* his dominant *guṇa* because that is what will ensure his overwhelming feeling of pleasure. This pleasure has to be distinguished from the sensual pleasure that we ordinarily speak of which is a result of gratification of worldly desires and are only of a transitory nature.

It is only for purposes of analysis that we have so far maintained a strict dualism between spirit and matter, between the sentient *Puruṣa* and the insentient *Prakṛti.* However, it should never be overlooked that they always exist side by side. One does not function without the other. The treatment so far dwelt exclusively on the various facets of *Prakṛti*, but we will now turn our attention to a discussion of *Puruṣa.* Without the discussion of this complementary aspect, the Sāṅkhya doctrine would lose all its force since it is likely to be mistaken for an outdated philosophy of nature, thus stripping it of all its true philosophical content. The existence of *Puruṣa* is arrived at by what is, in modern terminology, called the argument from design which we have discussed in chapter 1. We said that although the design argument is suspect from a scientific perspective, its importance in philosophical discussions cannot be denied. In all arguments from design, one recognizes a predetermined architecture of the necessary apparatus for achieving a transcendental purpose. In the Sāṅkhya doctrine, the existence of the *Puruṣa* is deduced as the transcendental principle governing the elaborate architecture of *Prakṛti.* The doctrine asserts that the unfolding of the primary evolutes of *Prakṛti* and the rich diversity of nature that emanates from *Prakṛti* has the sole purpose of serving the transcendent *Puruṣa.*

We now reiterate the two Sāṅkhya doctrines, one relating to *Prakṛti* and a second relating to *Puruṣa.* The doctrine governing the justification of *Prakṛti* as an ultimate reality of matter is derived from *satkāryavāda*, the argument from effect to the first cause. The doctrine governing the justification of *Puruṣa* as an ultimate reality of the spirit is based on the argument from design, that is arguing from a regular pattern of things to their final cause.

The usual interpretation of Sāṅkhya philosophy as regards *Puruṣa* is that of pluralism because the *Puruṣas* are manifold, which accounts for the individuality of humans. In fact, this viewpoint assumes the proportions of a major criticism from those adhering to the philosophy of non-dualism in which the identity of the cosmic principle and the psychic principle is its foremost message. However, the protagonists of the Sāṅkhya philosophy ward off this criticism by saying that the plurality of *Puruṣas* should be understood by the analogy of the integral relationship that exists between the innumerable rays of the sun and the sun. The assertion is therefore that there is only one *Puruṣa* in the sense that the rays of the sun, corresponding to the several *Puruṣas*, have no separate existence from the sun.

According to them, the identity of the individual *Puruṣa* to the Cosmic *Puruṣa* is quite evident in the doctrine without any need for further elaboration.

Puruṣa is static by nature. The internal senses are absolutely necessary for *Puruṣa* to know or will. While *Puruṣa* is quintessentially sentient in nature, its association with *Prakṛti* is absolutely necessary for the experience of psychic life. The feelings of enjoyer (*bhoktṛ*) and witness (*sakṣi*) would be possible to experience only when *Puruṣa* works in association with the internal senses.

Is it ever possible for spirit to be completely independent from matter, or what is the same, for *Puruṣa* to be isolated from *Prakṛti*? We said earlier that the internal senses are the permanent accompaniment of *Puruṣa* even after the death of the gross body. The definition of the subtle body in terms of the internal senses affords a clear distinction from the gross body. But it is only when the internal senses become latent that there is complete freedom for *Puruṣa* from *Prakṛti*. Such a state is indeed recognized and is called *kaivalya*. This is a stage which is attained only when there is no rebirth, since birth always means the reappearance of a unique set of internal senses, the uniqueness understood in terms of the degree of perfection of the *sattva guṇa*, depending on one's own spiritual evolution. Such a stage corresponds to the end of *sámsāra* of the *Karma* doctrine. *Kaivalya*, therefore, refers to the final stage of self-realization.

The various philosophical schools indulge in detailed discussions, of the *pramāṇas*, which are the valid means of acquiring knowledge that are unique to them, and also deal with their theses on perceptual knowledge. The Sāṅkhya school is no exception to this philosophical tradition. The purpose, of course, is to establish as rational a thesis as possible on the basis of the central concepts of the school, which in this case are *Puruṣa* and *Prakṛti* with all their ramifications, in order to provide the intellectual insights about the transcendental reality. A detailed understanding of relative existence based on the primary evolutes of *Prakṛti* is considered essential for projecting the limits of rational thinking out to the transcendental realm. Thus, the Sāṅkhya philosophy is an argument proceeding from the relative realm of worldly realities towards the absolute realm of the transcendental.

The argument about subjective and objective reality that we discussed earlier is common to both science and philosophy. What we mean by subjective reality is that whenever an external object is apprehended, there is no uniform perception on the part of everyone apprehending it. To put it in common language, one always looks at the world through one's own rose-colored spectacles. Because all the characteristics of an object are never perceived, individuals will see it differently. This is what explains the diversity of world views. The objective view, on the other hand, does not contain this ambiguity. It is the kind of truth contained in the statement that the world exists and will go on forever irrespective of our individual perceptions. It is a worldly reality experienced by everyone without dispute.

The understanding of subjective reality is explained in terms of the Sāṅkhya vocabulary. The internal senses, as we know, are the permanent accompaniment of *Puruṣa* and therefore exist without beginning in time. Furthermore, they are unique to an individual. Perception of an external object by the sentient *Puruṣa*, which is static by nature, is always present when it acts in conjunction with the internal senses, which are dynamic in nature. It is this mechanism of perception of external reality which accounts for its subjectivism. The view that objects are perceived indirectly through the media of the internal senses with some semblance of reality is called the *theory of representative perception*. It is opposed to the view of perception which says that objects are known directly.

Sāṅkhya, however, attaches equal importance to both aspects of reality. If an overemphasis is laid on the subjective aspect, one can only expect distortions of truth because, by its very nature, it can never reveal the total truth in all its aspects. The assertion about giving equal importance to both subjective and objective reality may therefore seem like a contradiction when we know definitely that the internal senses will never permit perception of total objectivity because of their past history which indeed is beginningless. To explain this philosophical riddle, we again appeal to the Sāṅkhya schematic. We note that the internal senses are essentially *sattvic* in nature, which means that their natural propensity is to be in a condition which would disallow subjectivism in the perception of truth. However, their contamination is due to an accumulation of the *rajas* and *tamas* elements which, in turn, are due to the past spiritual history of the individual. The contamination occurs because of the impressions which are left on the subtle body, the *saṁskāras*, in the process of thought and action. It is believed that these *saṁskāras* are not only accumulated in this life, but also inherited from previous births. That is why we say that the past history of the internal senses is beginningless. The defects in the internal senses due to *rajas* and *tamas* in what is essentially *sattvic* are to be overcome through spiritual practice. Since the three *guṇas* always coexist, what one can at best hope to achieve is to minimize the influence of the two *guṇas* of *rajas* and *tamas* on *sattva guṇa* by reducing their proportions and ensuring their favorable interactions. We are assured that this process will always ensure the dominance of the *sattva guṇa* which, for all intents and purposes, will enable one to apprehend the objective reality in all its totality. In short, it is spiritual evolution which will enable one to progressively overcome the handicap of subjective perception resulting in an incapacity to perceive an external object in its true colors.

Spiritual evolution is, in the Sāṅkhya schematic, tantamount to climbing the inverted tree all the way towards the evolute of *Buddhi* and beyond. When this ascent is complete, one gains the intuitive knowledge to be able to distinguish *Prakṛti* from *Puruṣa*. This state is called *vivekajñāna* and is the be-all and end-all of all philosophical thought. The differentiated consciousness of the individual will merge into the infinite consciousness of the *Puruṣa*, thus ending all the feeling of separability.

As a slight digression, we refer back to an earlier discussion of chapter 4 where we discussed the Omega Point theory due to the French theologian Teilhard de Chardin. There, we alluded to the fact that a correspondence exists between the concept of the Omega Point and the concept of *mahat* of *Vedānta*. In our discussion of *Prakṛti*, we have called its first evolute as Buddhi which is only an alternate name to *mahat*. Just as the Omega Point, *mahat* has not only an evolutive aspect in the relative field, but also a transcendent aspect in the realm of the absolute. The significance of this analogy between the concepts of the Omega Point and *mahat* can be further explored because of the current scientific interest in the subject as in *The Physics of Immortality* [45].

The intuitive knowledge one gains at the apex of the Sāṅkhya schematic is totally different from secular knowledge which is obtained within the realm of rationality. Secular knowledge could encompass a detailed understanding of nature gained through a study of the various disciplines which explain its diverse phenomena by using theories such as the theory of relativity, quantum mechanics, Darwin's theory of biological evolution, or the big bang theory of cosmology. Intuitive knowledge, on the other hand, is not an aggregate knowledge. It is the synthetic knowledge obtained by training the mind to experience the consciousness of the infinite as exemplified by the *Puruṣa* rather than knowledge obtained through a study of the external world. No access to a library is necessary for obtaining intuitive knowledge since its requirements are based on an inward spiritual odyssey. It is not even an intellectual pursuit and so is also accessible for those who are not intellectually inclined. Reaching such a state is considered to be the ultimate goal of life. It should never be overlooked that the physical aids *Prakṛti* provides, which are the mind and body of the human being, are absolutely indispensable for gaining intuitive knowledge for the perception of the unsublatable reality which is transcendental in nature. That is why so much emphasis is also laid on the physical welfare of the human body and the sanity of the mind.

The Sāṅkhya philosophy further explains what constitutes knowledge by giving a more detailed account of the mechanism of perception. We have already stated that both *Puruṣa* and the internal senses are necessary for perception to take place. This composite of spirit and matter, of *Puruṣa* and internal senses is called *Jīva* (soul), and knowledge can only arise when there is a modification of *Jīva*. The modification, of course, is only in the internal senses since *Puruṣa*, by definition, is always without change. When an object is to be perceived, the internal senses shoot out like a dart and assume the form of the object under consideration. Put simply, the common experience is that there is perception only if we set our mind on the object of perception whether it be through the medium of seeing, hearing, tasting etc. When the mind assumes the particular form of the object, the unique mode of the internal senses, which is called *vr̀tti*, is illumined by *Puruṣa*, the static element of *Jīva*, and it is this state which is called the knowledge of the object. Obviously, the modes keep changing depending upon the objects of perception. If there were no such changes, there would be constant perception or constant non-perception, which we know are absurd propositions. Even the knowledge of internal feelings such as pleasure and pain are explained on the basis of this model of perception.

What about the experience of *kaivalya* which is declared as the main aim of life? How does this model of perception explain that state of mind? We said that knowledge results as a modification of the internal senses, and in the case of their total aloofness from *Prakṛti*, which is what *kaivalya* is, there is no evidence of internal senses at all. Consequently, the question of knowledge as we normally understand it does not arise. *Kaivalya* is therefore an experience, albeit of a very special kind, without any knowledge of the environment as such.

The Sāṅkhya doctrine is purely a theoretical philosophical construct without any explicit reference to its practical aspects. In the course of our discussion, we made statements pertaining to spiritual practice and also to the attainment of the ultimate goal of *kaivalya*. But there is nothing said about the actual steps to be taken by a spiritual aspirant in order to tread a practical path. These essential concepts are explained in the philosophy of Yoga which is paired with the Sāṅkhya school.

5.2 Yoga

The Yoga philosophy due to sage Patanjali is described in 195 aphorisms and is paired with sage Kapila's Sāṅkhya philosophy. Sāṅkhya and Yoga are considered to be the two facets of *satkāryavāda*, the former emphasizing the theoretical aspect of the doctrine and the latter its practical aspect. The Yoga philosophy, although complete in itself, does not explicitly indulge in metaphysical speculations about

Man and the universe. Rather, it diagnoses what the fundamental spiritual ailment of humanity is and proceeds to offer a permanent cure for it in a manner analogous to what a medical doctor would do in identifying a physical disease afflicting a patient and proceeds with the administration of the required therapy for the restoration of health. This clinical approach for dealing with the spiritual problem of the human race is contained in four statements:

- Life is full of *duḥkha* (sorrow);
- The cause of *duḥkha* is spiritual ignorance (*avidyā*);
- The eradication of *duḥkha* is entirely possible; and
- There is a definite way for the eradication of *duḥkha*.

The above four statements are usually referred to by the four words which rhyme in Sanskrit, *heya*, *hetu*, *han* and *hanopaya*.

In passing, it is interesting to note that very similar statements appear in the enunciation of Buddhist philosophy, which is one of the reasons the Sāṅkhya philosophers claim the imprint of their doctrines on Buddhism. These are the famous Four Noble Truths, which are:

- Life is full of sorrow;
- The cause of sorrow and suffering is due to ignorance;
- The eradication of sorrow and suffering is entirely possible; and lastly,
- The remedy consists in following the Noble Eight-fold Path.

The first statement, *heya*, is a simple declaration about the malady of the human spirit which is found in prince and pauper alike. The experience of sorrow is considered to be only a symptom of this underlying disease. The second statement, *hetu* refers to the fundamental cause of spiritual ignorance (*avidyā*) which is what propels one to seek permanent happiness in the world of our ordinary experience. The third statement, *han*, injects a note of optimism that there is a definite cure for this ailment. Without this assurance, the philosophy would have been utterly pessimistic. The fourth statement, *hanopaya* consists of a detailed prescription for the restoration of normal spiritual health which is the main kernel of the Yoga philosophy.

In passing, it is interesting to note what His Holiness John Paul II has to say about the Buddhist tradition in his book *Crossing The Threshold Of Hope* [32]. He says that the methods derived from that tradition have a "*negative soteriology.*" At two different points in the text, the Pope says:

> The " enlightenment" experienced by Buddha comes down to the conviction that the world is bad, that it is the source of evil and suffering for man. To liberate oneself from this evil, one must free oneself from this world, necessitating a break with the ties that join us to external reality–ties existing in our human nature, in our psyche, in our bodies. The more we are liberated from these ties, the more we become indifferent to what is in the world, and the more we are freed from suffering, from the evil that has its source in the world... To indulge in a negative attitude toward the world, in the conviction that it is only a source of suffering for man and that he therefore must break away from it is fundamentally contrary to the development of both man himself and the world, which the Creator has given and entrusted to man as his task.

The reason why the above comments from the vicar of Christ on Buddhism become pertinent to our study is because they are indirectly applicable to some of the doctrines of Hinduism also. For instance, we have traced some commonality in the initial statements of the philosophies of Yoga and Buddhism. There is a lesson to be learnt from these comments. Since the followers of the traditions of Hinduism and Buddhism do not subscribe to a pessimistic view of life, a *negative soteriology*, it becomes incumbent upon those expounding these philosophies to highlight the positive aspects to a sufficient degree as to avoid all possible misinterpretations. It is well-known that inter-religious debates are fraught with difficulties and so one can only try to put forth one's point of view in the clearest of terms taking the rival criticisms into account in the hope that some day in the future there will be a better understanding between all the religions of the world.

We have stated earlier that the Sāṅkhya philosophy is criticized as atheistic in character because of the absence of the specific mention of God in the exposition of its doctrine. Even in Yoga philosophy, which is its natural pair, the mention of God does not appear in a prominent way. However, the importance of this philosophy in the clear articulation of the basic human problem as it appears in the four statements and the practical means for overcoming it is uniformly appreciated by all other schools of philosophy.

There may be a rational explanation from the philosophy of science as to why sage Patanjali does not indulge in an esoteric discussion of the metaphysical problems concerning man and the universe, but instead concentrates on the practical aspects only. A clue to this understanding is contained in the 'Operational Philosophy' championed by the Nobel Laureate physicist Percy Bridgman [4]. The main premise of this philosophy of science is that a question is meaningful only when there is a way for answering it; one must be able to identify an intellectual operation, on the basis of which one could adequately answer the question. Statements which cannot be subjected to verification on the basis of corresponding intellectual operations are considered *neither true nor false.* Bridgman was able to demonstrate the efficacy of his philosophy in the discipline of physics. He did it in such a way that it provided deep insights for its validation in many other scientific disciplines.

We shall now examine the implications of Operational Philosophy to philosophical questions. The operations in this case are not intellectual, as in the case of science, but they can be regarded as mystical or spiritual operations. Sage Patanjali's step-by-step development of the Yoga philosophy can be understood on the basis of Operational Philosophy which consists of operations applicable to the physical, moral, mental and spiritual realms. We have called these mystical operations with the final aim of the philosophy in mind, but they can be further subdivided into the particular realms of their application. The Yoga philosophy is expounded with such great precision that it is possible to clearly identify the mystical operation for the verification of the experience of *samādhi*, which is an important step to be achieved as a means for unleashing the contrary knowledge to overcome spiritual ignorance.

In this philosophy, however, the assertion about God does not have a well-defined set of mystical operations which could serve the purpose of its verification. It is important, however, to reiterate that its non-verification in terms of mystical operations only means that the assertion about God is neither true nor untrue. This conclusion does not amount to atheism. Viewed from this perspective of Operational Philosophy applied to the philosophical realm, one has to be careful before leveling the charge of atheism to the Sāṅkhya–Yoga pair of philosophies. It is interesting in this context to note what Aldous Huxley had to say in his book *Grey Eminence* [25] to explain the criticism of atheism leveled against Buddha. He invokes the argument of Operational Philosophy based on mystical operations to explain why Buddha, while giving specific answers to questions on Liberation and Enlightenment, completely avoids answering questions about God. The latter could not be be answered on the basis of Operational Philosophy, while the former concepts were very much within its purview.

The word Yoga, in common usage, is used in a very restrictive sense to connote some physical postures of the body whose repetitive use helps the practitioner gain better health. Without minimizing its importance, one should note that this aspect is only a very small facet of the Yoga philosophy which is complete in itself as an independent treatise. Not only does it provide the practical counterpart to the Sāṅkhya doctrine, but it also wields a profound influence on all other systems of Indian philosophy. Of late, Yoga is referred as a secular discipline in order to suggest that its practice does not conflict with other religious faiths. We shall only make a brief reference to its tenets because much of it is best understood while in actual practice under the guidance of a qualified master. Also, it is not in our outline to include a detailed discussion of it in this book.

One of the significant topics of the Yoga philosophy is *aṣṭānga-yoga*, which means the eight limbs, *angas*, which are its accessories. The eight limbs stated in order are self-restraint, *yama*; observance, *niyama*; posture, *āsana*; regulation of breath, *prāṇāyāma*; withdrawal of the senses, *pratyāhāra*; steadying of the mind, *dharana*; contemplation, *dhyāna*; and meditative trance, *samādhi*. These eight steps are normally viewed in terms of a staircase model where one climbs step by step from the bottom to the top.

We shall now very briefly revert to the classical explanation of *aṣṭānga-yoga*. The first two steps are meant for the purposes of moral and ethical cleansing and they contain lists of do's and dont's. The first limb, *Yama*, is a list of the negative aspects. These include non-injury, *ahiṁsa*; adherence to truth, *satya*; avoidance of tendencies to steal other's property, *asteya*; celibacy, *brahmacarya*; and disowning possessions, *aparigraha*.

The second limb is *Niyama* which emphasizes the positive things one has to achieve: purity, *sauca*; contentment, *saṁtoṣa*; right aspiration, *tapas*; study of philosophical texts, *syādhyāya*; and devotion to God, *Īsvara-praṇidhāna*. Note that this is the first place where God appears in our entire discussion of the Sāṅkhya–Yoga school. The discussion on Sāṅkhya was completely devoid of any such reference, and as a result, the entire doctrine is severely criticized as being atheistic. It is only when the Sāṅkhya-Yoga dual pair is considered together that this criticism gets muted because the Patanjali doctrine recognizes God, though not in a prominent way. However, practitioners of the Sāṅkhya system consider this criticism as totally misplaced because for them, the personification of the concept of *Puruṣa* does indeed provide them with the devotional appeal of a personal God.

The assiduous practice of *Yama* and *Niyama* is said to result in a mental state of dispassion , *vairāgya*, which is the natural tendency to distance oneself from worldly desires, while at the same time, very much participating in worldly affairs. The emphasis is on the inward transformation resulting from the conquest of the ego factor and not on a superficial outward behavior mimicking the state of dispassion. This moral strength is considered to be the prerequisite for the ascent to the next step. The next three steps of *āsana*, *prāṇāyāma* and *pratyāhāra* are meant to take control of the bodily functions necessary for spiritual practice.

It is the *āsana* aspect which is commonly referred to as Yoga where the sole emphasis is on the fitness of the body only, thus distorting its true meaning as a means for spiritual perfection. There is an extensive literature on *āsanas*, and it is best learnt under practical guidance as are the next two steps.

Prāṇāyāma works on the principle that *prāṇa*, vital air, and *prajñā*, consciousness, are interrelated. We observe, for instance, that the rate of breathing goes down as the tranquillity of the mind increases, and conversely. It is a mind-body relationship of immense significance which we discussed in chapter 4. This relationship has gained recognition in the current medical literature which emphasizes the importance of holistic medicine. *Prāṇāyāma*, which can be roughly translated as breathing exercises, though very specifically defined ones, is practiced in order to gain the ability to still the mind, which is the main purpose of meditation. It is interesting to note that the practice of meditation will automatically reduce the rate of breathing thus lending credence to the feedback model of the eight limbs which we will discuss next. The practical aspects are best learnt under the guidance of a qualified teacher.

The next three limbs, namely, *dhāraṇā*, *dhyāna* and *samādhi* are the final steps which have to be practiced for gaining mastery of the mind. It has to be emphasized that *samādhi*, which means meditative trance, is not the final goal of the Yoga doctrine. It is only the means for realizing the Sāṅkhya–Yoga truth.

We shall now present another interpretation of the Yoga of the eight limbs advocated by teachers who take a rather dim view of the interpretation given by the staircase model because, in their understanding, it is almost impossible to satisfy the complete requirements of one step before proceeding to the next. In other words, it would be too much to ask that a spiritual aspirant complete all the requirements of self-restraint, which is the first limb of *aṣṭāṅga-yoga*, before proceeding to the practice of the next limb of observance, and so on until reaching the finals steps of *dhyāna* (meditation) and *samādhi*. These teachers point out such an involved and long drawn out procedure becomes so painful that the final goal of this practical philosophy may turn out to be beyond reach.

The alternative interpretation they provide is that the eight limbs, instead of being viewed as in the staircase model, should be viewed as overlapping spheres where the experience of the succeeding stages, however fleeting they may be, influence the cultivation of the previous stages. We shall refer to this as the *feedback model*. It is characterized by the fact that part of the output of the seventh sphere on meditation is fed back to all other preceding spheres, as in all control mechanisms. For instance, a modicum of experience in meditation will most certainly enhance the possibility of cultivation of self-restraint which is the first step. Not only that, this possibility is what makes the whole exercise pleasant instead of making it agonizingly painful without the end in sight. Furthermore, progress in this Yogic discipline is measured by the simultaneous achievement of the different sets of goals prescribed for each sphere. It is interesting to note that the exponents of the feedback model also claim that Sage Patanjali's thesis admits of such an interpretation. As a matter of practical detail, preceptors of *aṣṭānga-yoga* do initiate a disciple into the meditative discipline at an early stage, which further reinforces the interpretation of the feedback model.

There are two stages of realization resulting from the Yogic discipline. The first is *samprajñatā*, where one remains conscious even when one realizes the intuitive truth of *Puruṣa*. When this stage is transcended, one attains the state of *asamprajñatā*, where *Puruṣa* exists with all its effulgence. When all internal senses are suspended, one becomes a *jīvanmukta*, at which point, the cycle of births and deaths ends even though when one is alive.

The Sāṅkhya-Yoga pair, when taken together, presents a complete picture of both the theoretical and practical aspects of the philosophy. Although this system is very old and its impact is felt on other systems of philosophy, in practice, it is not, however, as popular as the philosophies which come under the *Upaniṣadic* schools. The reasons for this can best be discussed after our discussion of the *advaita* philosophy due to Śaṁkara. For the present, we shall conclude the chapter by recounting some of the oft-repeated criticisms that we have already encountered at several places during the course of our discussion.

In the Sāṅkhya-Yoga philosophical school, God is not explicitly mentioned except as one of the aims of the second limb, namely *niyama*, where *Īśvara* is postulated. Sāṅkhya, by itself, is atheistic since there is no explicit mention of God in that doctrine. The concept of *Puruṣa* is established on the basis of an argument from design, but it never makes mention of a designer. The concept of God of the Yoga school is merely to ensure the evolution of *Prakṛti* unlike the *Upaniṣadic* God which arises from the Brahman–Ātman concept. Devotion to God is not central to the exposition. The duality of the static *Puruṣa* and dynamic *Prakṛti* leaves some questions unanswered about the role of the insentient *Prakṛti* being vested with the dynamic power through the command structure of its evolutes. There is also some question about the arbitrary designations of the evolutes. There is almost the topsy-turvy implication that it is nature that makes the decision for the salvation of the *Puruṣa.*

The criticisms just stated are, of course, refuted with vigor by ardent followers of the Sāṅkhya school. They are particularly incensed by the criticism about the singular absence of the concept of God in their doctrine. For them the personification of *Puruṣa* is indeed their concept of a personal God which is very much there for anyone who cares to see. In their opinion, there is no extra justification required to formulate the identity of the psychic principle with the cosmic principle because it is implicit in the philosophy. First, *Puruṣa* and *Prakṛti* always coexist, and so the ultimate reality of matter is never divorced from the spiritual element. Secondly, there is no distinction made between the cosmic *Puruṣa* and the individual *Puruṣa,* following the analogy of the sun and its rays. Last but not least, the Yoga doctrine, which is a natural pair to the Sāṅkhya doctrine, is an extensive thesis of all the practical aspects which almost every other philosophical school embraces.

Chapter 6

Śaṁkara's Nondualism: (Advaita)

Having discussed Sāṅkhya, we shall present a brief account of the philosophical school of non-dualism (*advaita*) due to Śaṁkara, or Śaṁkarācārya as he is commonly referred to with great reverence. In fact, the treatment of chapters 2, 3 and 4 is based on this branch of Vedantic philosophy. This *Upaniṣadic* school of philosophy is called ***absolutistic*** in contrast to the theistic schools which are also based on the *Upaniṣads.* Since these latter texts appear at the end of the Vedas, they are also called *Vedāntic* schools.

Before we proceed with the study of the Śaṁkara school, it would be useful to recount some of the main philosophical issues that were subjects of considerable controversy, as well as the seminal contribution of Badarāyaṇa (also called Veda Vyāsa) that provided the much-needed intellectual clarity and coherence to spiritual thought. We already stated in chapter 2 that this great sage was a prodigious scholar and a spiritual giant in whose memory India observes a national holiday called *guru pūrṇima.* The sheer volume of his lasting contributions to scriptural literature provides just cause for the unstinting praise that is heaped on him. He is considered to have systematized the vast Vedic literature which was, until that time, being transmitted orally in a rather disorganized fashion. He did so by organizing the vast array of material into texts on the four Vedas, with their subsections. He is believed to be the author of the *Bhāgavata* literature which has the purpose of conveying the central message of the Vedas in an expansive style in order to make them more accessible to the general public. The great epic of *Mahābhārata,* which includes the sacred text of Gītā that is accorded the same importance as the *Upaniṣads*, is also attributed to him. Finally, pertinent to our present discussion, we recall that he was the author of *Vedānta Sūtras*, also called *Brahma Sūtras*, which were intended to allay the confusion surrounding the meaning that had to be attached to some of the key statements (*mahāvākyas*) of the *Upaniṣads*.

Vyāsa's principal objectives in writing the *Vedānta Sūtras* were manifold, some of which were fulfilled with great clarity and some others not so because of the terse style of the text. He is believed to have successfully refuted the dualistic interpretation of the Sāṅkhya school which also claims its authenticity on the *Upaniṣadic* texts. In that doctrine, for all its analytical elegance and comprehensive coverage of philosophical concepts, the interrelationship between the dual concepts of the sentient Puruṣa and the insentient *Prakṛti* was never considered to be satisfactorily explained. Also, Vyāsa disproved the claim of the *Mīmāṁsa* school that the significance of *Karma Kāṇda* should be accorded more prominence than the message of the *Upaniṣads*. The former emphasizes the importance of sacrifices and rituals, and in the process, completely soft pedals the importance assigned to spiritual knowledge, which is the main theme of the latter. Vyāsa thus restored the true vision of the *Upaniṣads* by correcting the topsy-turvy view of the *Mīmāṁsa* school.

One other contribution of *Vedānta Sūtras* that is of significance to Śaṁkara's philosophy was the discussion about the nature of the relationship between the absolute *Brahman* and the universe. During the time of Vyāsa, even amongst those who considered *Brahman* as the absolute, the prevailing view was that the universe arose from *Brahman*, which is called *Brahma pariṇāmavāda*. This was in contrast to the philosophy of Sāṅkhya which considers *Prakṛti* as the first cause for the creation of the universe and as such called it *Prakṛti pariṇāmavāda*. Vyāsa is believed to have differed from the *Brahma pariṇāmavāda* school and instead upheld the view that the universe is only a phenomenal appearance of *Brahman*, in the same sense that the dream state of consciousness is only so when viewed from the vantage point of the ordinary waking state of consciousness. This latter doctrine, which was later attested by Śaṁkara, is called *vivarta vāda* and differs from the law of causality called *satkāryavāda* that we encountered earlier.

It is important to note that not all *Vedāntic* schools subscribe to the absolutistic interpretation of the *Upaniṣads*, and as such they do not necessarily uphold either *Brahma pariṇāmavāda* or *vivarta vāda*. However, all of them state their philosophical positions by giving their own commentaries on *Vedānta Sūtras*. But we will confine our discussion to Śaṁkara's philosophy of *advaita* not only because of its great popularity, but also because its sheer majesty in philosophical thinking provides ample scope for drawing on the insights from scientific thinking.

As we commented earlier on, maintenance of accurate historical records was not one of the strong points of Indian genius with the consequence that Indians are invariably forced to talk about the philosophical doctrines of several sages without ever including any biographical data about them. Thus, access is denied to a vital source of information which would enrich all philosophical discussion by introducing the human element. This stands in stark contrast to the western tradition of maintaining faithful records of the life histories of all its great proponents. This glaring lapse is sometimes explained away on the basis of the lack of importance Indian philosophical thinking attached to secular matters in general, but that somehow smacks of a poor excuse. In any case, many do not regard it as a convincing explanation on the basis of some contrary evidence. It cannot be overlooked that Indian scholars were very eager and supportive of the research done by the orientalists in digging into the historicity of the scriptures and even now are keen on claiming the glory of Vedic heritage and the great contributions the country has made in all the diverse disciplines. One only ought to sense the pride that contemporary Indians attach to what has come to be known as Vedic mathematics. Against this background of paucity of historic information about spiritual masters, the availability of a biographical sketch of Śaṁkara, however sketchy it may be, comes as a great relief since it gives a human touch to his rather abstract philosophical thesis.

Śaṁkara was a man of titanic intelligence, a spiritual giant, and perhaps the greatest missionary of Hinduism. Unfortunately, no accurate historical records have been maintained of even this prodigal son of India whose monumental contributions to Hinduism have endured even to this day. However, scholars have put together some brief sketches of his life and times through indirect sources whose authenticity has generally been accepted. His date and place of birth have been put at 788 A.D. in the village of Kalady on the banks of the river Periyār in the state of Kerala which is at the southern most tip of India. What is truly amazing is that he completed his most productive mission in the cause of restoring the true vision of Hinduism within a very short life span of 32 years.

Śaṁkara made his appearance at a time of total spiritual decadence of the country. The great spiritual message of Lord Buddha that had provided a true insight into the nature of the ultimate Reality had in course of time been totally distorted by his followers which is not an uncommon occurrence in the history of great religions. This was, at least, the Hindu perspective on the Buddhism of the times. As for the Vedic message, it was completely skewed towards the strict ritualistic observance of the religion which is the main purport of the *Karma Kāṇda* portion of the Vedas. This practice was done almost to the exclusion of the later section on *Upaniṣadic* knowledge, thus robbing the Vedas of all their meaning and substance. All manner of teachers had sprung up in different parts of the country, resulting in a dissonance of views on the real purport of the Vedas. Thus, the time was ripe for Śaṁkara to restore the true vision of the Vedic philosophy by removing all the cobwebs of misunderstanding surrounding it. This he did with supreme success from the lofty platform of his own Self-realization. His phenomenal intellectual prowess and unrivaled ability for engaging the protagonists of other views in debate are recounted with great admiration in many historical accounts. It is often said that Śaṁkara drove Buddhism out of India. This statement, suggesting a sense of triumph, however, should be understood in its proper perspective. The Hindus have always held Buddha in the highest regard, and so there could never have been any question of hostility towards Buddhism. When Śaṁkara convincingly established his new vision of an integrated life based on his interpretation of the Vedas, particularly of the section on *Upaniṣads*, it made it impossible for Buddhism to prevail in the country although it spread to the neighboring countries and very soon became an acknowledged world religion in its own right.

There was no political concept of a well-knit state during the time of Śaṁkara, with the result that there was no unifying force to pull all the different regions of the country together. During a period where the modes of transportation were primitive, Śaṁkara travelled tirelessly throughout the length and breadth of the country to spread his ageless spiritual message which is called *sanātana dharma.* This is not an organized religion, rather it is a message addressed to every individual reminding him of his spiritual heritage on the basis of which he could lead a purposeful life with robust optimism. In course of time, the message took firm roots in the country. Also, it provided the real solidifying force for a country of such vast cultural diversity. Śaṁkara established centers of learning, four monasteries *(mutts)* in all the four corners of the country to ensure the propagation of his philosophy in all its pristine purity from generation to generation. This in itself is an amazing feat considering the fact that proper modes of transportation were practically nonexistent in that vast country. One special feature of these mutts that he established is their tradition of maintaining an uninterrupted lineage of teachers and disciples, called *guru-śiṣya paramparā.* It is truly astonishing to note that such lineage have been preserved from the eighth century onwards even to this day. We end this short biographical note on Śaṁkara with a quotation from Sister Niveditā who was a western disciple of Swami Vivekānanda [13]: In devotion he was like St. Francis of Assissi; in intellect he was like Abelard; in dynamism and freedom, he was like Martin Luther; in imagination and efficiency, he was like Ignatius Loyola. In fact, he was all these, united and exemplified in one person.

We stated earlier that sage Veda Vyāsa wrote his *Vedānta Sūtras*, also called *Brahma Sūtras*, with a view to clarifying some of the principal statements that appear in the *Upaniṣads* regarding the relationship between *Brahman* and *ātman*. Unfortunately, this treatise is also written in a cryptic style thus adding some ambiguities of its own to the confusions in the interpretations that already prevailed of the supposed meaning of the original texts. Vyāsa's purpose in writing his thesis is, however, considered to be very clear on some of the major contentious points. As stated earlier, he wanted to refute the dualistic interpretation of the ancient Sāṅkhya doctrine, which was very popular in his days. He also wanted to establish the primacy of place for the *Upaniṣads*, which appear in the concluding sections of the Vedas, in view of the prevailing claims of the *Mīmāṁsa* doctrine which had given almost exclusive prominence to religious rites and sacrifices that appear in the earlier *Karma kāṇda* portion of the Vedas. Lastly, he wanted to discredit the thesis that the universe emerges from *Brahman*, a thesis known as *Brahma pariṇāmavāda*. It is difficult to assign the order of importance he attached to the three major philosophical positions, although some scholars of the Śaṁkara school would rank the criticism of *Brahma parināmavāda* as the most significant one for reasons which will become clear later.

Śaṁkara wrote his own commentary on the *Vedānta Sūtras*, which was almost mandatory for any philosopher of repute for establishing his *bona fides* amongst the *Vedāntic* schools. By doing so, he clearly established his own school of philosophy called *advaita*, which means non-dualism. His thesis is called Absolutistic since he views *Brahman* as the Absolute. This terminology helps us to distinguish it from other theistic schools of *Vedānta* where the ultimate truth is regarded as one's own personal God.

We shall first briefly mention some of Śaṁkara's principal ideas on the interrelationship between spirit and matter on one hand, and spirit and the empirical self on the other. He considers the universe as a phenomenal appearance of *Brahman* which is called *vivarta vāda*. This is in contrast to the prevailing idea of *Brahma pariṇāmavāda* which includes the assertion that the universe is a manifestation of *Brahman*. Śaṁkara also proposed the *Māyā* doctrine in order to explain the relationship between *Brahman* and the universe. Consistent with this doctrine, he also enunciated his view about the relationship between *Brahman* and the empirical self (*jīva*). His conception of spiritual liberation (*mokṣa*) was based on the merging of *jīva* with *Brahman* after *jīva* divests itself of the constraint of spiritual ignorance (*avidyā*) imposed on it. He was very firm in his pronouncement that spiritual ignorance can only be overcome through its contrary knowledge called *jñāna.* This was in contrast to the *Mīmāṁsa* doctrine where the emphasis was on action rather than knowledge. The religious dimension to the *advaita* doctrine is contained in the concept of *Saguṇa Brahman* (qualified *Brahman*) which ensures the theistic ideal of the *Upaniṣadic* God. And lastly, he was meticulous in his exposition of the concept of *Brahman*, also referred to as *Nirguṇa Brahman* (*Brahman* without qualifications) in order to highlight the differences in its concept from both *Saguṇa Brahman* and the Buddhist concept of void (*śūnya*). We shall amplify the treatment of these concepts as we proceed with a discussion of the *advaita* thesis.

The metaphysical problem is two-fold: it deals with the relationship of the spirit with the physical universe on one hand and the relationship of spirit with the individual soul on the other. The paradoxical conjunction of the physical and psychical components in the conception of the empirical self is quite central to the philosophical problem. In *advaita*, the psychical component is called *sākṣin*, the eternal witness, and it can be equated to the concept of *puruṣa* of Sāṅkhya. The different philosophical schools sometimes have the same word to connote different meanings, and so it becomes necessary to understand at the very outset what the technical words mean in a specific study. Similarly, at the level of the cosmos as a whole, there is a conceptual similarity in details between *Prakṛti* of Sāṅkhya and *Māyā* of *advaita*, although there is a basic difference in their conceptualization. We have stated these approximate equivalencies between the central concepts of *advaita* and Sāṅkhya in order to portray some of the similarities between the two doctrines. The major differences between them, however, arise when we discuss the manner in which the advaitic doctrine deals with the metaphysical problem.

We have already said that Śaṁkara's interpretation of the *Upaniṣads* is classified as absolutistic. There can be several explanations for this, but for the sake of continuity, we will deal with one based on our earlier discussion of Sāṅkhya philosophy in the previous chapter. In Sāṅkhya, we find a clear separation made between the realms of matter and spirit through the dual concepts of *Prakṛti* and *Puruṣa* which are accorded equal importance. It was pointed out that the ultimate goal of a spiritual aspirant was *kaivalya*, which meant escaping from the bondage of *Prakṛti* in order to merge with *Puruṣa.* However, the criticism leveled against the Sāṅkhya doctrine is that the relationship between the dual concepts of *Puruṣa* and *Prakṛti* is never reconciled satisfactorily. Essentially, the doctrine unfolds from the relative field of existence and proceeds towards the absolute. The interplay of the three *guṇas* and the dynamics of the hierarchy of evolutes were both subject to the law of causality that is operative in the relative field of existence in which the three-fold requisites of space, time, and causality come into the picture. In view of these philosophical underpinnings, Sāṅkhya is classified as dualistic in nature.

In contrast to the dualism of Sāṅkhya, Śaṁkara recognizes *Brahman* as the only reality. In the chapter on Vedic philosophy, we pointed out that the key statements of the *Upaniṣads* were taken to mean that the inner essence of the empirical self, which is the psychic principle of *ātman*, is identical to the cosmic principle of *Brahman* governing the universe. *Ātman* is the true essence for the feeling of constancy that we experience in the 'I' notion prevalent in the paradoxical conjunction of being and becoming. Śaṁkara's thesis emphatically asserts that *Brahman* is the only unsublatable reality and that all philosophical discussions should proceed from that premise, which is why his philosophy is called Absolutistic. All his deliberations pertaining to the interrelationships between spirit and matter on one hand, and spirit and the empirical self on the other, proceed from this lofty platform of the Absolute. The thesis is based on the premise that we have to explore the relationship of the relativistic field of existence on the basis of what is already known. On the other hand, exploring the ultimate reality of the universe from some premise within the relative field of existence would amount to exploring the unknown from the unknown. This shaky premise is described by the metaphor, 'searching for a black cat in pitch-darkness particularly when it is not there'. The absolutistic thesis avoids this pitfall and proceeds from what is definitely known on the basis of Vedic testimony, which is the psychic principle at the level of the individual self and its identity with the cosmic principle. The discussion therefore proceeds from this transcendental certainty.

It is interesting to grasp the force of this philosophical argument in light of the insights provided by Gödel's theorem in mathematics. We commented earlier that the philosophical implication of that theorem was that we can never successfully explore the infinite from the realm of the finite. Śaṁkara's emphatic assertion that the philosophical investigation should proceed from the Absolute rather than from considerations of the relative field of existence can be viewed as a qualitative assertion of the famous mathematical theorem which has implications for disciplines other than mathematics. (See Penrose [31].) If it did not proceed in this manner, it would result in the hopeless situation of first investing the unknown universe with an ultimate reality and then proceeding to investigate the existence of this reality while placing oneself in the realm of the unknown.

One should recognize that the law of causality does not hold in the realm of the Absolute. Time, space and causality have meaning only in our comprehension of worldly realities. As discussed in chapter 1, these concepts have validity only after the creation of the universe. Accordingly, Śaṁkara's philosophy differs from the premises of Sāṅkhya's *Prakṛti pariṇāmavāda* and the then-prevailing absolutistic version called *Brahma pariṇāmavāda*, which are both dependent on the notion of cause-and-effect relationships, though of different kinds. In the former it refers to the cause-and-effect relations arising from *Prakṛti* and its evolutes with no internal contradiction, since they all refer to the relative field of existence. However, Śaṁkara argued that the *Brahma pariṇāmavāda* philosophy has a serious conceptual error: while it correctly recognizes *Brahman* as the Absolute, as in *advaita*, it commits the mistake of invoking a cause-and-effect relationship for establishing the relationship between *Brahman* and the universe by declaring that the universe is a manifestation of *Brahman*. It is inconceivable to think of a cause-and-effect relationship between the eternity of the Absolute and the concept of time, which gets embedded from the moment of creation of the universe. Consequently, one can expect that Śaṁkara's *advaita* steers clear of cause-and-effect relations in establishing the relation between *Brahman* and the physical world.

The main contributions of Śaṁkara's *advaita* philosophy can now be stated before filling in the details. The *Māyā* doctrine is proposed to serve as the link between the Absolute and the physical universe, between eternity and time. The doctrine is careful to preserve the integrity of the two domains of the Absolute and the Relative with reference to notions of space, time and causality. This is achieved by declaring the universe as only a phenomenal appearance of *Brahman* rather than its manifestation. The identity between the Absolute and the empirical self (*jīva*) comes into play when the limiting adjuncts on *jīva* are not operative. These are the same adjuncts caused by spiritual ignorance which we discussed earlier in the last chapter. The union of *jīva* with *Brahman* is the ultimate spiritual release, a state called *mokṣa*. *Mokṣa* is a concept which arises on the basis of this absolutistic doctrine, whereas the Sāṅkhya concept of *kaivalya* arose when the discussion proceeded from the relative to the absolute. Of immense significance is the optimistic assertion that *mokṣa*, which is the state of liberation, can be achieved in one's present life. The limiting adjunct on *jīva* is one's own spiritual ignorance which is called *avidyā*. The way to nullify the effects of *avidyā* is by gaining the contrary knowledge called *jñāna*.

The emphasis here is that spiritual knowledge alone is the ultimate liberating factor. The fulfillment of the moral and ethical injunctions to cleanse the doors of perception, which is the primary emphasis of the *Mīmāṁsa* philosophy, serves only as a means towards achieving liberation; it is not an end in itself. The concept of *Saguṇa Brahman* which appears in the thesis is the exact philosophical counterpart of *Īśvara* which is the personal God of religion. In order to understand the necessity for this concept of *Brahman* with a prefix, one should recognize that God and devotion, that are essential parts of religion, are in the field of relativity, and as such the corresponding infinite consciousness operating in that field should be distinguished from the pure consciousness of the Absolute. In order to clearly maintain the difference, the latter is called *Nirguṇa Brahman* which means *Brahman* without attributes. The *Saguṇa Brahman* acquires all the divinely opulent attributes of *Īśvara.* This interpretation is essential for the practice of religion within Śaṁkara's Absolutistic doctrine. Also, Śaṁkara found it necessary to point out the dissimilarities between his concept of *Nirguṇa Brahman* and the Buddhist concept of *śūnya*, meaning 'void'. The latter concept has connotations of being negativistic and nihilistic, whereas the former does not even remotely suggest that description. Articulation of this difference was deemed important because of the seeming similarity between the two concepts particularly since the Buddhistic influence was very prevalent at that time. We shall not, however, refer to this later in our discussion.

In Śaṁkara's thesis, the concept of illusion (*mithyā*) takes a center stage in the discussion. This prominence becomes necessary in order to explain the two-fold relationships between *Brahman* and the universe, and *Brahman* and *jīva.* These two relationships are explained on the basis of two types of illusions.

We shall first comment on the illusion associated with the relationship between *Brahman* and the universe. One classical example that is often cited in order to explain what is meant by illusion in this context is the case where one mistakes a rope for a snake. This illusion will last as long as the misperception lasts. As soon as the correct knowledge of the rope dawns, the illusion of seeing a snake in the place of a rope will suddenly vanish. This example provides an useful analogy for considering the superposition of the world of diversity on the unchanging *Brahman.* As soon as the correct knowledge of *Brahman* is realized, the illusion of the reality of the universe will suddenly disappear.

It is the concept of cosmic illusion (*māyā*) that provides the linkage between the Absolute and the physical world. We have repeatedly emphasized that the transcendental field of the Absolute is beyond the reach of space, time and causality, while the universe is very much described by those three prerequisites. The concept of illusion caused by the superposition of the two realms will not vitiate the requirement that the link should preserve the integrity of the two realms with respect to the three prerequisites. Figuratively speaking, the link should not have one node in the Absolute and the other in the relativistic field of the universe. Such, indeed, was the fallacy of *Brahma pariṇāmavdāa*, which as Śaṁkara argued, was a flawed version of the Absolutistic *Vedānta.* The concept of illusion completely obviates the need for an appeal to the law of causality based on cause-and-effect relationships, as in *satkāryavāda.* The phenomenal relationship is instead based in *vivarta vāda.* It has to be emphasized that due to spiritual ignorance, the illusion is experienced only in the relativistic field of our worldly realities. The illusion is suddenly eradicated once the contrary knowledge to overcome the ignorance dawns on us. In summary, the relationship between Brahman and the world is explained on the basis of an illusion. In view of this, the world is sometimes referred to as a dependent reality or an apparent reality, which is the essence of *vivarta vāda.*

Next, we discuss the relationship between Brahman and the empirical self (*Jīva*) which is explained on the basis of a second type of illusion. In the first type of illusion, that we just discussed, the entity which is superposing the real one is totally fictitious. In the illusion encountered in the snake and rope example, the snake was a figment of the perceiver's imagination. In the second type of illusion that we are now discussing, the object of illusion is very much existent, only the perceiver is registering a distorted image of it. A familiar illustration of this is one viewing the world through one's own rose-colored spectacles. It is important to realize that, in this example, the object is very much there and that the termination of illusion will reveal it in its true colors. More specifically, the illusion is not caused by a mental state and so cannot be described as subjective.

The classical example that is given to reinforce the idea of this second type of illusion is the case of a white conch viewed through the medium of a yellow glass. It is assumed that the viewer is blissfully unaware of the presence of the interposing yellow glass. Under those conditions, one mistakes the white conch to be yellow. When knowledge of this distortion dawns on the viewer, he immediately recognizes that the yellow color of the conch is due to the obstructing medium, and hence a property of the conch is immediately correctly perceived. The presence of the conch is not denied; only its yellowness, which was perceived as an aspect of it, is denied. The sheet of glass is called an adjunct (*upādhi*), which is externally imposed on the white conch.

Based on the above analogy, the relation between *Brahman* and *Jīva* can be explained. *Jīva* is really *Brahman* itself but for the constraints imposed on it. The adjuncts associated with *jīva* are the internal senses which are *Buddhi*, *Ahaṁkāra*, and *Manas* (*antaḥkaraṇa*), which we have already discussed in connection with the Sāṅkhya schematic. Since they are evolutes of the physical world, they are also physical in their characterization. Both the interposition of the adjuncts and their eventual removal are entirely within the physical world. The *Brahman* is untouched by either the cause of the illusion or its eventual dissolution.

We can now conclude on the basis of the two illusions, one operating at the level of the universe at large and the other at the level of the individual, that *Brahman*, which is one without a second, appears both as the physical world and as the empirical self. In the first case, the *ultimate reality* of the physical world is totally denied. And in the second, only the physical adjuncts associated with *Jīva* are denied. With the removal of these adjuncts, the spiritual element present in *Jīva* is identified as *Brahman* itself. Putting it in the language of being and becoming, which are the twin aspects of constancy and change that are present in the empirical self, it is the three modes of mind which are identified by the internal senses that are responsible for the state of becoming; these three are the limiting adjuncts and with the removal of their influence, the residual spiritual component of the composite entity of *Jīva* is recognized to be none other than *ātman*, which is identical to *Brahman*. Accordingly, *Jīva* is not at all false or illusory, which is an obvious conclusion based on our own experiences of consciousness and self-awareness. In fact, from our discussion of the *Karma* doctrine, we know that *Jīva* is considered beginning-less (*anādi*).

We shall further elaborate on the four central concepts of Śaṁkara's philosophy of *advaita*. These are 1) *Māyā* 2) *Brahman* 3) *Saguṇa-Brahman*, and 4) *Jīva*. In this list of four concepts, it is only *Māyā* that is a physical entity while the other three are spiritual entities.

Māyā is identical in conception to *Prakṛti* of Sāṅkhya with respect to its description of nature and its evolutes. The Sāṅkhya schematic could very well be called the schematic of *Māyā* as far as the portrayal of nature's diversity is concerned. The major differences between the two concepts arise only when their respective relationships with *Puruṣa* and *Brahman* are considered. Sāṅkhya suggests the relationship between the two incompatibles of spirit and matter by appealing to ordinary human experience which vouches for their coexistence and inseparability at all times. In the example, 'I am typing', the 'I' refers to the spiritual element and the word 'typing' refers to the physical act. *Advaita* vehemently denies that there could be any such direct relationship between spirit and matter and invokes the concept of illusion to explain that relationship. Since the spiritual element is the only reality, the appearance is attributed to spiritual ignorance which acts as a limiting adjunct. An extension of this philosophical reasoning based on analogy asserts that there is a similar relationship between *Brahman* and the physical world where the latter is only a phenomenal appearance of the former due to *Māyā*. The physical world therefore is regarded as neither real nor unreal. *Māyā* has two aspects to it. The first is the veiling influence (*āvaraṇa śakti*), by virtue of which it effectively screens out the knowledge of *Brahman*. The second is the diversifying power (*vikṣepa śakti*) which is responsible for the phenomenal world of names and forms. It is this which characterizes the rich diversity of the universe.

As for the concept of *Brahman*, it is the substratum, the divine Ground of both the physical world and the empirical self. In chapter 1, we considered the laws of nature as representing the element of constancy corresponding to the component of being at the level of the cosmos. We did so based on our observation of regularity and dependability of nature which is what provides the motivation for scientists to discover laws exhibiting the features of constancy. While part of the scientific enterprise is directed towards the unification of such laws to bring in more and more coherence, there is a concurrent effort aimed at investigating the diversity of nature. There seems to be general agreement in the scientific community that there is no limit to these two activities since nature is 'infinite in all directions'. When we say that *Brahman* is the substratum of the physical world, we are also asserting that the cosmic principle underlying the universe is spiritual in nature because of its identity with the psychic principle of *ātman*. The validity for this assertion is based on the principal declarations (*mahāvākyas*) of the *Upaniṣads*. The primary and secondary evolutes of *Māyā* represent actual changes in *Māyā* in the process of its manifestation. The changes, however, are only virtual with respect to *Brahman*. They are merely appearances because of the cosmic illusion implied by the superposition of the physical world on the Absolute. *Brahman* remains changeless for ever.

As for establishing the identity of *Jīva* with *Brahman*, Śaṁkara's doctrine is careful in not considering the latter as an object of knowledge since subject-object relations are not valid in the transcendental field. The identification of the spiritual component of the empirical self with *Brahman* removes all ambiguity. The 'I' of all experiences, which is the element of constancy in the empirical self, is found to be the psychic principle which is the same as *Brahman*.

Next we discuss the concept of saguṇa *Brahman* (qualified *Brahman*) whose justification is based on an intricate philosophical argument. The purpose for evoking this concept is, however, to recognize a level of consciousness that exactly corresponds to the God of our universe which is called *Īśvara*. We can expect to reach out to the Lord only in the world that we are living in. Consequently, the corresponding consciousness cannot be *Brahman*, which is pure consciousness in the transcendental field. In order to arrive at the concept of *saguṇa Brahman*, we note that both *Brahman* and *Māyā* are causative factors for the universe although in very different senses. In the former, the cause is apparent while in the latter the cause is actual; the evolutes of *Māyā* are definitely based on cause-and-effect relationships defined by *satkāryavāda*. It is a combination of these two principles of causation, one that is apparent and another which is actual, that gives rise to the notion of *Saguṇa Brahman*.

This philosophical concept can be understood in contrast to the Sāṅkhya doctrine of *Prakṛti* where the latter becomes the source of all creation and evolution without any explicit appeal to *Puruṣa*, albeit that it always coexists with *Prakṛti*. *Saguṇa Brahman* brings together both spirit and matter in a unique way by avoiding the ambiguity attributed to the Sāṅkhya doctrine. In order to understand this rather involved relationship between spirit and matter, it is important to refer to the concept of time in the *Advaita* thesis, since that is what is central to the law of causality. Time is considered as a relation between spirit and *Māyā*; we again remind ourselves that *māyā* is a phenomenal concept since it cannot be said that it exists, nor can it be said that it does not exist. it neither exists nor exists. Since the spirit is the Reality and *Māyā* is only an appearance, their relationship signified by time is also considered as phenomenal.

In contrast to time, the concept of space arises only after the event of creation and therefore it comes within the domain of time. It is thus considered to be entirely an offshoot of *Māyā* since it arises immediately after the 'moment of creation', which is an idea that conforms to with scientific thinking. Consequently, in the philosophical thinking of *advaita*, while time is phenomenal, space is an evolute of the physical universe, and as such the two are on an entirely different footing. Space presupposes that the principle of causation has already begun to sprout, whereas time precedes all such conception. Space is in time only, unlike all other entities of the world which are both in space and time. All these involved concepts and much more are at the basis of the conceptualization of *Saguṇa Brahman*. The easiest way to regard *Saguṇa Brahman* is that it is the philosophical counterpart of the theistic ideal of Īśvara. The relationship between *Saguṇa Brahman* and the physical universe is unique.

Saguṇa Brahman is, therefore, the philosophical concept arising out of the combined considerations of *Brahman* and *Māyā*. However, a discussion of the law of causality apropos to the realms of the Absolute and the Relative, and the corresponding conception of the personal God of *Īśvara* is now of interest to us. In fact, *Īśvara* has the universal appeal of a religion with its stress on love and devotion. The dual concept of the devotee and his personal God is more tangible for the majority of people than the more seemingly abstract non-dual concept implied by the identity of the *jīva* with *Brahman*. A careful examination of the adjuncts we have associated with *Īśvara* and *Jīva* will reveal that this *Upaniṣadic* God will not be held responsible for all the pain and tragedy we witness in this world. The locus of *māyā* which is *Īśvara*'s adjunct is the same as that of *Brahman*, and so it is also infinite. Consequently, *Īśvara* remains untouched by the changes in mental states of the individual self. Such changes in mental states are caused by the finite adjunct of *avidyā*. In this conception of God, He is the creator of the universe, and *māyā* is His divine power (*śakti*). God is also referred to as *māyāvi* meaning wielder of *māyā*, a magician in the figurative sense that He performs the great magic of creation without ever being deluded by the grand spectacle. *Īśvara* is both the efficient and material cause for this universe. He is the efficient cause because He is the creator, and He is the material cause because it is *Māyā* that represents the undifferentiated matter. The God of the *Upaniṣads*, therefore, is different in conception from the God of the Semitic religions where God is considered just as the creator.

The close correspondence between the philosophical concept of *Saguṇa Brahman* and the religious concept of *Īśvara* paves the way for couching the central essence of the *advaitic* thesis in purely religious terms. We can say that the ultimate purpose of religion is to completely surrender the ego, the 'I' notion, through pure devotion (*bhakti*) at the feet of *Īṣvara*. The act of such supreme devotion to the Lord through which one loses one's personal identity is indeed equivalent to gaining the knowledge that is necessary to overcome spiritual ignorance in order to achieve the merger of *Jīva* with *Brahman*. This type of devotion wherein the devotee does not ask for rewards from his personal God, not even liberation for that matter, is very different from the more common case of petitionary prayers where the goal is to ask for some type of immediate relief or the other. The lofty type of devotion is equivalent in all respects to the path of attaining the highest type of knowledge of the Self. It is in this respect that knowledge and devotion become indistinguishable.

Jīva is a hybrid entity consisting of both the spiritual element, namely the Self, and an insentient element due to the internal senses; it describes the coexistence of being and becoming. The aspect of becoming is ascribed to spiritual ignorance (*avidyā*). Since the philosophical construct has throughout assumed a parallel between the microcosm of the individual and the macrocosm of the universe, we can say that *avidyā* is the component of *māyā* at the individual level. The characteristic feature of *Jīva* is that it wrongly identifies itself with the gross body because of the influence of the internal senses that have the natural propensity to project outwards rather than inwards towards the sentient element. This wrong identification is the root cause of all the discomfiture in life. It should be noted that this built-in illusion inherent in the empirical ego precedes all that one experiences in life. Spiritual liberation consists of overcoming the primary ignorance of wrong identification. In contrast to this, the cosmic illusion of *Māyā* does not inject any confusion in the conception of God because there is no wrong identification of any sort at the macrocosmic level.

What remains when the *avidyā* aspect is completely obliterated is the pure sentient element which is called *sākṣi*, the eternal witness. Sakṣin is the equivalent of *Puruṣa* in the Sāṅkhya philosophy. It is also called the pure consciousness (*svarūpa jñāna*).

Advaita also believes in representative perception, just as in Sāṅkhya. The composite entity that is responsible for perception consists of *sākṣi* and the internal senses. It is the latter that undergoes incessant changes in order to perceive objects, whether they are external to the body or internal to it at the level of feelings. These are called modifications of the internal senses which are responsible for giving rise to knowledge of various kinds. During all of this process, *sākṣi*, which is the sentient element, remains without any change. Perception occurs when the internal senses interpose between *sākṣi* and the object. The mediating senses are supposed to assume the form of the external object when it is seized by an external sense. The particular form determined by the object is called its 'mode' (*vṛtti*). When the knowledge of such a mode is illumined by the *sākṣi*, it results in *vṛtti jñāna*. A similar argument holds with respect to internal feelings like pleasure and pain. In such cases, the external senses do not come into the picture. The internal senses undergo suitable modifications for enabling direct perception.

In conclusion, perception is dependent upon both the changeless *sākṣi* and the changes that occur in the internal senses, called their modes, which are defined by the objects of perception.

A discussion on knowledge and error will lend further clarity to the central tenets of *advaita* which explain the relation between *Brahman* and the physical world on the basis of *Māyā*, and the relation between *Brahman* and *Jīva* on the basis of a second type of illusion.

Broadly speaking, there are two distinct philosophical views of knowledge that can be identified. The first is the familiar view that knowledge always points to an object outside. The second is the view that there is no object apart from its knowledge; the distinctions that are made between knowledge and its contents is a figment of one's own imagination. (We have discussed these differing concepts of reality in section 4.4.) Very often, we come across contradictory positions taken on this classification of knowledge. Insistence on the existence of an object when knowledge is true and its denial when knowledge is in error constitutes such an inconsistency. Śaṁkara takes an unequivocal stand on what constitutes knowledge and avoids all such incongruities. According to him, all knowledge is predicated upon not only the existence of a corresponding object but also a perceiving subject. It is impossible to conceive of a knowledge which does not have the twin implication of a subject and an object. Conversely, if an object does not exist, there can be no question of its corresponding knowledge. The classical example, a rather tasteless one from our present-day sensitivities, that is cited in support of this observation is a barren woman's son. We will assume for purposes of discussion that the impossibility suggested in this ancient example is still valid regardless of the tremendous possibilities of modern day genetic technology! The son of a barren woman exists only in verbal jargon. Since he is entirely fictitious, there can be no knowledge corresponding to this entity.

When such a categorical stand is taken on the relationship between object and knowledge, one is faced with the burden of explaining how illusions occur. *Advaita* makes the assumption that there is knowledge even in the case of an illusory object. It only makes a distinction between illusory and non-illusory knowledge by recognizing that there is a basic difference between the two kinds of objects. For instance, in the example of the illusion caused by mistaking a rope for a snake, the illusory experience is confined only to the specific person experiencing it, and so it can be legitimately classified as a private experience. Furthermore, the private experience of a snake will last only in the interim period before the dawning of the correct knowledge. In contrast to this, the experience of a real serpent is an object of public or empirical knowledge (*vyavahārika*) as opposed to the private illusory knowledge (*pratibhāṣika*). It is important to note in this context that the private objects of one's illusion are not deemed subjective in nature. They are, in fact, objects of the mind of the illusory experience.

Very often Śaṁkara's philosophy is criticized because of the concept of illusion, which is an integral part of it. Some critics consider the concept as totally redundant while others totally distort its true meaning and make derogatory references to it in their polemical arguments against *advaita*. In order to clear such misunderstandings, we again emphasize that *advaita* holds worldly realities as an illusion only when they are viewed from the realm of the transcendental field. This viewpoint is sadly mistaken to mean that the worldly realities are an illusion, thus stripping the statement of its real meaning. But Śaṁkara's critique, in which even illusory experiences are explained on the basis of objective knowledge, should put an end to all such ill-considered criticism.

The error which is implicit in illusory experience is attributed to an 'illegitimate transference' (*adhyāsa*) of the false knowledge onto the real one. While empirical and apparent knowledge are both objective, they are however, classified into two different categories of objects exemplifying two different grades of reality. In general, this can be called two orders of being, one apparent and the other empirical. In the example of the rope and snake, the apparent knowledge of the snake represents one order, and the real knowledge of the rope refers to a different order. Obviously, the apparent order has less reality attached to it than the empirical reality. The confusion caused by the superposition of the two orders of reality comes to an end as soon as the enduring order comes to the fore. *Adhyāsa*, therefore, can be defined as the confusion induced by superposition of the two orders of being and is caused by the ignorance surrounding it. This ignorance (*avidyā*) has two aspects to it: first, there is the concealment (*āvaraṇa*) of the true order of reality, which is the rope in the example under discussion; secondly, it projects the false knowledge of the apparent order (*vikṣepa*) by inducing the illusion of a snake in the place of the rope. This ignorance is suddenly eradicated when the contrary knowledge (*vidyā*) arises. We have already discussed the concept of illusion caused by (*avidyā*) at the individual level. The ultimate source of this ignorance is *Māyā*, which is operating at the cosmic level.

The phenomenon of the individual ego (*jīva*) can now be analyzed on the basis of the concept of *adhyāsa*. We said earlier that *jīva* consists of the spiritual element, which is the *sākṣi*, and the physical elements corresponding to the internal senses (*anthakarna*). These two entities are also responsible for the two juxtaposed concepts of being and becoming at the level of the individual ego. In *jīva*, we find both the aspects of subject and object. The *adhyāsa* in the case of *jīva* has as its basis the ignorance caused by the two orders of being, namely, the *sākṣi* and the internal senses. In the example 'I am working', the working is due to the experience of the internal senses and not due to *sākṣi*.

If we also take the transcendental reality of *sākṣi* into account, we can consider three orders of being: the first is due to *sākṣi*; the second is the empirical reality due to the internal senses which is *avidyā*; and the third is the apparent reality. We have already seen that error is introduced whenever the apparent reality (*pratibhāṣika*) is mistaken for empirical reality (*vyavahārika).* In this context, truth implies that one should relate objects of the same order. However, we should be cautious to limit this definition to the empirical order only. This is because the objects of a dream, though belonging to the same order, get sublated immediately on termination of the dream. This happens because both the subject and object of experience in a dream vanish once one wakes up.

Based on the above discussion, a valid means of acquiring knowledge (*pramāṇa*) is defined as that which leads to true knowledge in the sense that it is not invalidated by any later knowledge. We say that what is less real is superposed on what is real. The error caused by the apparent knowledge is suddenly erased the moment the empirical knowledge dawns on the subject.

It would be fruitless to search for criteria for knowledge of the absolute from within the relative field of existence characterized by diversity. While it is true that there is a relentless search for unifying principles in any discipline, they can, at best, serve the purpose of bringing more and more *coherence* to the individual disciplines. The discipline of cosmology is a case in point. It has resulted in magnificent advances in bringing together the physics of the universe at large, which is the theory of relativity, with the particle physics of quantum theory. Interesting theories are advanced for the big bang phenomenon, and yet at the same time, the theories are careful to state that they can only explain what happened to the universe from the 'moment of creation'. But the world of the absolute, as we know, is before creation where the definition of time is not even applicable. This is why the concept of the moment of creation becomes ambiguous and needs more sophisticated mathematical concepts to explain it. From this point of view, mere coherence of the physical theories, impressive as they are when viewed within the domain of physics, can never be a distinguishing feature of knowledge of the absolute. It is easy to see that we will come to similar conclusions on the basis of any other discipline. In each of the disciplines, one can look for coherence on the basis of their respective intellectual origins. Hence, in philosophical terms, we say that the criterion for distinguishing knowledge of the absolute should be *comprehensive* in nature. It is so because all divisions of knowledge emanate from the undifferentiated consciousness.

We conclude this discussion about criteria for knowledge by referring to the concept of infinity in mathematics, for purposes of lending further clarity to the aforesaid discussion. Since the discipline of mathematics, which incidentally, is so closely linked to the investigation of the laws of nature, embraces the concept of infinity, the question may be asked whether it is capable of reaching out to the absolute, which is also characterized by infinity. This important question was answered in the negative by the famous British mathematician George Cantor whose name is associated with set theory. His conclusion is based on some properties which a set of infinities have to satisfy; we guardedly say that there is more than one infinity in mathematics in order to distinguish them from the contexts in which these arise. For instance, there is the infinite set of integers, the infinite set of fractions, and so on. It turns out that the set of infinite sets does not satisfy the property of a set, and consequently, the concept of infinity of mathematics cannot embrace the concept of the absolute, which is infinite in character. The important conclusion is that even mathematics is of no help in comprehending the infinity of the absolute.

In summary, no criteria for knowledge of the absolute can be determined within the field of diversity. Since the absolute is beyond the reach of the limits of rational thought, one would expect this conclusion to be true. *The criterion for knowledge of the absolute is comprehensiveness since it is the real source of all other knowledge of the phenomenal universe. The criteria of correspondence, meaning correspondence of knowledge with an object, and cohesion, both of which are applicable to knowledge of the relative field of existence are not enough to qualify as distinguishing features of knowledge of the absolute.*

The goal of human life is implied in the discussion of *jīva*. Since the spiritual element in this composite entity is nothing other than pure consciousness but for the constraints operating on it, the goal of human life should therefore be to try and eradicate the influence of the constraints. Under the influence of spiritual ignorance, which is the limiting adjunct on the sentient element, one invariably has a tendency to wrongly identify oneself with the gross body. The only way to overcome spiritual ignorance is through its contrary knowledge which is called *jñāna*. The change that is expected of a spiritual aspirant is an attitudinal one rather than one based on real change. Knowledge of *Brahman* has not to be sought afresh because it is already there in self-luminous form. Only the ignorance surrounding it has to be ended. The change that is expected of a spiritual aspirant is therefore virtual rather than actual.

Professor Hiriyanna, in his scholarly book on Indian philosophy, *The Essentials of Indian philosophy* [20], gives a very interesting example to illustrate this idea. He compares the lunar eclipse with the solar eclipse in order to distinguish actual change from virtual change. In lunar eclipse, where the moon is obscured by the shadow of the earth, clear vision is restored only when the moon's position with respect to the sun and the earth undergoes an actual change. In contrast to this, in the case of the solar eclipse, the sun, which is the luminary, remains in the same position, and it is the position of the observer with respect to the sun and the moon that is the cause of the eclipse. In this case, the removal of obscuration of the sun asks only for a virtual change. Based on this analogy, we conclude that *Brahman*, which is the luminary, is changeless and all that is expected of the spiritual aspirant is to acquire the correct knowledge of it by shifting his focus away from the influence of the external world.

Although the goal of spiritual realization can be clearly stated, its achievement will, however, call for the fulfillment of some prerequisites which demand a life-long dedication to the goal. First, it is important to acquire right knowledge about the essence of the Vedic philosophy. But mere intellectual knowledge alone will not suffice. The second prerequisite is to undergo the disciplines that are stipulated for moral and ethical cleansing. For the acquisition of right knowledge, the three stage process of *śravaṇa, manana* and *nididhyāsana* is recommended. These steps are hearing about the ultimate truth from a qualified preceptor (*guru*), constant contemplation on the truth that is instructed, and finally, direct experience through meditation. We have already commented on these three aspects of acquiring right knowledge in chapter three. As for the disciplines that have to be rigorously pursued in order to achieve mental purification, these refer to various types of actions that have to be performed and are best learnt in the course of religious practice. These latter aspects of religious rites and moral duties are considered absolutely necessary, and they are complementary to the procedures for acquiring right knowledge, though secondary in importance within the total scheme of things.

All these steps might appear very involved and suggest a long and tedious journey. But, in practice, all these phases overlap with each other and in the process lend more and more efficacy to the practice of the discipline. The central point here is that the philosophy concerning the reality of *Brahman* need not be relegated to the certainty implied by a theological dogma. Scriptural knowledge can at best serve only as a compass, and beyond that what is expected is actual realization of the truth that is proclaimed. Once it is achieved, even the support of the scriptures is rendered unnecessary.

Mokṣa is the state of spiritual liberation that is attained when the disciple realizes the ultimate truth. Since *Brahman* can not be an object of knowledge, realization means the experience of total identity with it which is the ultimate non-dual experience. One who is so liberated is called a *jīvanmukata.* He, who has achieved this supreme state of plenary consciousness, would have put an end to the endless cycle of births and deaths that is ordained in the *Karma* doctrine. Such a person can continue to lead a purposeful worldly life with robust optimism. Living in eternal bliss, he is no longer tempted by the cravings of ordinary mortals. The pains and pleasures of the world no longer leave permanent impressions on him. There is strong evidence to suggest that his spontaneous thoughts and actions would not only serve as a source of inspiration to the rest of humanity, but also result in positive good for society as a whole. There is a widely prevalent belief amongst the Hindus that India is progressing despite incalculable odds because the country has never lost its constant supply of sages who are silently radiating their divine energies for the benefit of mankind.

Chapter 7

Theistic Schools: Rāmānuja and Madhva

Having presented Śaṁkara's absolutistic interpretation of *Vedānta*, we now proceed to a theistic interpretation of *Vedānta* due to Rāmānuja who came almost three centuries later. Rāmānuja's philosophy is called *Viśiṣṭādvaita*, and according to popular understanding, it takes a midway position between the two extremes of non-dualism and dualism pertaining to the two relations between *Brahman* and *Jīva* on one hand, and *Brahman* and *Prakṛti* on the other. We have seen that in Śaṁkara's philosophy of non-dualism these two relations are explained in terms of an identity based on the illusory nature of *Prakṛti*, both at the level of the cosmos as a whole and at the level of the finite self. The dualistic thesis, which we will discuss later in this chapter, asserts the ultimate reality of both Jīva and *Prakṛti* and summarily dismisses the argument of non-dualism while explaining their relations with *Brahman*. The principal exponent of this dualistic thesis is Mādhvācārya (1199–1278 A.D.) who came after Rāmānuja. It is important to note that all these *Vedāntic* schools claim to be authentic interpretations of the *Upaniṣadic* texts and appeal to *Vedānta Sūtras* by Bādarāyaṇa for their legitimacy.

Rāmānuja was by no means the first to come up with the *viśiṣṭādvaita* thesis. He was preceded by other eminent teachers, the most prominent of whom was Nāthamuni. Also, he was succeeded two centuries later by the most versatile teacher Venkatanātha (1350 A.D.), who acquired the saintly name of *Vedāntadeśika*. His prodigious scholarly contributions resulted in greater consolidation of this school of philosophy. However, the principal credit for the systemization of the *viśiṣṭādvaita* thesis goes to Rāmānuja because of his unrivaled genius in accomplishing the synthesis of several diverse currents of philosophical thought and religious practice.

There are several strands of thought that are intertwined in Rāmānuja's philosophy. First, he had to meet the demands of his practicing religion which was widely followed from ancient times. He belonged to the age-old devotional creed of Vaiṣṇavism, which is based on the patron deity (*Iṣṭadevatā*) of Lord Viṣṇu that had a long history in India. Vaiṣṇavism derives its authenticity from the *Upaniṣads* as well as its main source of the *Bhāgavatas*, and so there was a compelling necessity to integrate its main tenets into the new school of philosophy. Secondly, as we have already commented, the *Upaniṣadic* texts contain some seemingly conflicting declarations about the relation between *Brahman* on one hand, and Jīva and *Prakṛti* on the other. It was Rāmānuja's view that Śaṁkara's non-dualism had not done justice to the set of declarations that contradicted his view. In particular, Śaṁkara's *Māyā* doctrine was severely criticized because of the introduction of what was viewed as a needless abstraction unsupported by Vedic authority. Rāmānuja felt the need for a new harmonious explanation of the entire *Upaniṣadic* texts that would take care of the apparently contradictory views conveyed in its principal statements.

Thirdly, he felt the need for incorporating the theistic faith embodied in the rich devotional hymns that were composed by a galaxy of Tāmil saints called the Ālvārs. The most prominent amongst them was Nammālvār who composed *Tiruvaimoli* which consists of one thousand exquisite hymns of devotion. In fact, this literature, whose purport is in consonance with the Sanskrit scriptures, had come to be known as Tāmil Vedas. Rāmānuja, who was steeped in the tradition of the Tāmil saints, drew his inspiration in equal measure from this source also, so much so that his school of philosophy came to be known as *ubhaya vedānta* indicating thereby the equal importance assigned to both the Sanskrit and the Tāmil sources.

And last but not least, he was anxious to make his philosophic message accessible to everyone who wished to tread the spiritual path without regard to complex societal differences existing at that time. In fact, this tradition of egalitarianism was already prevalent, judging by the fact that some of the prominent Ālvārs were drawn from sections of the society which were not traditionally known for its high priests. Thus, we witness in Rāmānuja's message a strong galvanizing spiritual force born out of deep compassion towards his fellow man. Rāmānuja succeeded in ingeniously accomplishing all these diverse objectives by coming up with a superb synthesis of philosophical thought and religious practice.

7.1 A short sketch of Rāmānuja's biography

We have observed earlier that there is a general paucity of historic material about some of the most important leaders of India's philosophical and religious history. Against this background, it is interesting to include a short sketch of Rāmānuja's biography. We start the presentation of Rāmānuja's life and times by first referring to Nāthamuni. He was the first teacher (*ācārya*) of *viśiṣṭādvaita* and was born around 916 A.D. in Tāmil Nādu when the imperial *Colas* were ruling the country. Such information is not only relevant from the point of view of establishing Rāmānuja's family history but also from the vantage point of highlighting the disciplic succession of the school. Nāthamuni was reputed to be a great mystic and a scholar of considerable eminence. He was responsible for systematizing the 4000 hymns of the Ālvārs, called the *prabandhams*, which he popularized in the sacred shrines of Vaiṣṇavism. Although there is evidence to support the claim that he wrote some of the original works on *viśiṣṭādvaita*, unfortunately, they were all lost to posterity.

The next prominent successor of the school was Yāmunācārya who was a grandson of Nāthamuni. He was also a great mystic in the tradition of the Ālvārs and a reputable scholar of *Vedānta*. He wrote several texts; the one called *Siddhitraya* is devoted to the elucidation of the central tenets of the *viśiṣṭādvaita* philosophy. Yāmuna had an eminent disciple in his grandson Śrīśailapūrṇa who had settled down in the modern day Tirupati which is a well known shrine for Vaiṣṇavism. He had a sister called Kāntimati who was married to a great Vedic scholar called Asuri Keśavasomayaji. Rāmānuja was the son of this devout couple.

Rāmānuja was born in 1017 A.D. in the town of Śrīperambadur, which is close to Chennai (Madras). By the time of his birth, the influence of Buddhism had practically vanished from India, in contrast to Śaṁkara's time when it was very much a living force. From his scholarly father Rāmānuja learnt all the traditional practices of the Vedic religion. After he lost his father at the age of seventeen, he studied under a distinguished teacher called Yādavaprakāśa who had deviated from *advaita* and had established his own school of philosophy called 'identity in difference' (*bhedābheda*). It was during this time that the aging Yāmuna had a chance of meeting the young Rāmānuja in the Lord Varadarāja temple at Kānci in Tāmil Nādu, and he was struck both by his attractive physical appearance and his divine radiance. Yāmuna was so greatly impressed that he is supposed to have prayed in the temple asking for a boon from Lord Varada to ensure that Rāmānuja become the torch bearer of the *viśiṣṭādvaita* faith.

The studies under Yādavaprakāśa came to an abrupt halt when Rāmānuja started asserting his own independent interpretations of the scriptural texts. The story is told that the teacher became so envious of the intellectual and spiritual prowess of his disciple that he even devised a wicked plot to kill him under the pretext of taking him on a pilgrimage to Banares. Fortunately, Rāmānuja survived, and legend has it that his mysterious survival was due to the divine intervention of Lord Varada of Kānci. It was after this ghastly incident that Rāmānuja chose another preceptor called Tirukacci Nambi who was a direct disciple of Yamuna. From him, he is supposed to have learnt the following essential message underlying *viśiṣṭādvaita*: ' I am the supreme Truth, the Way and the Goal. The world of souls is different from Me as its source and sustenance. Self surrender is the way to salvation'.

From then on he donned the mantle of the *ācārya* of the tradition and became its authentic expositor. In order to fully devote his time to this cause, he became a renunciate (*sanyāsi*) and very soon acquired the reputation of being a prince of ascetics (*Yatirāja*). He made his home in the southern city of Śrīrangam where the Lord Ranganātha temple is situated. He devoted the remaining seventy years of his life to the cause of systematizing the philosophical school of *viśiṣṭādvaita* and spreading the gospel of the Śrīvaiṣṇva faith. He also brought a much-needed order to the Viṣṇu temples and effected their revival. Through his many writings, which of course, included commentaries on the *Brahma Sūtras* and the sacred text of Gītā, the new system was well-supplied with its own unique philosophical literature. He also produced what are known as *gadyas* which are characteristic for their outpourings of pure devotion.

Unfortunately, even a spiritual teacher of such eminence who did not entertain even a trace of sectarian hostility towards the followers of other systems of philosophical thought became a target for an inquisition from the reigning monarch of the day. The persecution of Śrivaiṣṇavism from the *Cola* king became so intense and so unbearable that Rāmānuja had to flee to the neighboring state of present day Karnātaka. The Hoysala kingdom of those days was ruled by a Jain king called Bittideva who had entertained a benevolent attitude towards faiths other than his own. When the king came into contact with Rāmānuja he also converted himself to the Śrīvaiṣṇava faith and assumed his new name of Viṣṇuvardhana. But his wife, Queen Śāntala, continued to adhere to her faith of Jainism; her doing so suggests that an enlightened attitude towards other faiths was prevalent in the kingdom.

Rāmānuja made Melkote, which is situated close to the city of Mysore, his seat of learning and paid a great deal of personal attention to the renovation of the Nārāyaṇa temple there. His approximately thirteen years of exodus from Tāmil

Nādu to Karnātaka turned out to be a period of renaissance for the Viṣṇu temples of the entire region, the most prominent amongst them being the Lord Keśava temple at Belur, which is now a tourist attraction because of its magnificent sculpture. Rāmānuja returned to Śrīrangam in 1118 A.D. after the death of the *Cola* king and continued his mission until his death in 1137 A.D. He is reputed to have been in full possession of his faculties till the last moment of his long span of life of 120 years. There are, of course, skeptics who doubt that he lived for so many years. They cite the dates for Nāhamuni and Yāmuna which are 823–923 A.D. and 916–1036 A.D., respectively as reason to question historical accuracy. One cannot deny that there is always the problem of establishing correct dates because of the poor tradition of maintaining accurate historical records, but there is ample evidence to support the facts that Rāmānuja did live to a ripe old age and that he relentlessly spread his gospel until his last breath.

7.2 The metaphysics of Viśiṣṭādvaita

As stated in chapter 1, the basic metaphysical question arises from the enigmatic conjunction of being and becoming at the level of the finite self. Starting from this premise, the philosophical inquiry within the *Vedāntic* context can be couched in more precise language when we take into account the psychic principle of *Ātman* and the cosmic principle of *Brahman*. The problem is to establish the relations that exist between *Brahman* and *Jīva* on one hand, and *Brahman* and *Prakṛti* on the other. All *Vedāntic* schools agree that the scriptural evidence for *Brahman* and *Ātman* is quite unambiguous but differ only on the question of their interrelationships. The differences arise because in some places the *Upaniṣadic* texts seem to suggest non-duality between the two phenomena, while in other places they appear to imply duality. Rāmānuja's masterly contribution was to solve this key puzzle by giving a new unequivocal interpretation to these seemingly contradictory statements of the *Upaniṣadic* texts. The non-dualistic interpretations had faced the criticism that the parts of the text that suggested dualism were glossed over in a contrived fashion, and the converse criticism was also true with respect to the later strictly dualistic interpretations. It was Rāmānuja's genius that addressed this difficult problem in order to restore total harmony to the interpretation of the entire text without encountering the criticism leveled against the two opposing schools of philosophy. But the differences voiced by the other two schools persist even to this day.

Rāmānuja's novel interpretation of the seemingly conflicting portions of the texts with regard to its key declarations such as *thou art that* takes us to a detailed examination of common linguistic usage when two distinct things are identified. His ultimate purpose was to establish the truths concerning the eternality and distinctiveness of the finite selves and the physical universe. When we say for example that *the rose is red*, we know that the rose is a substance and the redness is a quality, and the two are entirely distinct. Despite this knowledge of the distinctiveness between substance and quality, we combine them by the use of a linking verb ' is' as if the two are the same. We do so in conformity with usage. Another important example of immense consequence to his philosophy is the statement *I am a man* in which we combine spirit and matter. In this case the identity established is between two distinct substances, one of which is a spiritual element.

After observing these two types of usage, the next big step in the argument is to recognize the property of uniqueness in linguistic usage linking two distinct things. The uniqueness can be stated as follows: *The identicalness of two distinct things in ordinary linguistic usage is permitted only in two cases: first, between substance and attribute, and secondly, between body and soul* . The meaning of the uniqueness becomes clearer by examining a counterexample. We can consider the conjunction of the two distinct things man and computer in the usage, 'a man and his computer', which does not connote any identity whatsoever because it does not signify an identity between either substance and attribute or between spirit and matter. Considering again the second relation in the uniqueness between body and soul (*śarīra* and *śarīri*), we note that the two substances are not only inseparable, but there is also the additional factor of dependence of body on soul. The dependence implies that the body can exist only when the soul also exists and that knowledge of the body is predicated upon the knowledge of the soul at the same time. It is this body-soul relationship that is invoked to infer the correct meaning from key *Upaniṣadic* statements such as *thou art that.*

Rāmānuja also emphasizes that the soul–body relationship of *Brahman* with the two entities of finite selves and the physical world completely excludes the possibility of the imperfections of the latter two being transferred on to *Brahman.* This is so because the principle of unity of the complex whole retains the distinctive identity of all the three entities. Even from observation of the individual body– soul relationship, we know that the defects of the human body do not affect the soul. This understanding is further strengthened by the consideration that the doctrine of *karma*, which is accountable for the worldly imperfections, is relatively autonomous and does not directly involve the transcendent God. However it is subordinate to the will of God since He can bestow His divine Grace on the devotee at any time that He chooses.

The property of uniqueness implied in the two specific cases of linguistic usage is known as *apṛthak-siddhi* which means inseparability. Accordingly, the entire discussion on the identity of two distinct things is sometimes referred to as the *theory of inseparability* and is pivotal to the understanding of Rāmānuja's philosophy. One might wonder at the appropriateness of having Rāmānuja's philosophical discussion of reality originate from grammar, but such doubts will soon vanish when one realizes the richness of the implications. In any case, grammatical precision becomes necessary in order to understand the correct import of a principal *Upaniṣadic* declaration such as *that art thou* where 'that' refers to *Brahman* and 'thou' refers to the finite self.

There are some who have criticized Rāmānuja for invoking arguments from linguistic usage for solving philosophical problems. In this connection, it is interesting to take note of the extremely rich tradition of Russell and Wittgenstein who are recognized as two of the greatest western philosophers of the last century. Their chief contributions were to accord importance to the study of language in philosophical studies. An important task of philosophy is to provide unassailable means for analyzing the language used for expression of thoughts. Lending clarity to thoughts by removing all confusion associated with the use of language on the basis of well-established criteria is considered to be of prime importance. From this modern perspective, Ramanuja's thesis based on linguistic usage of the principal statements of Upaniṣadic texts does not seem odd at all. In fact, it is now deemed as a perfectly legitimate philosophical inquiry.

The relations between *Brahman* and the soul and between Brahman and the physical universe which constitute the crux of the metaphysical problem are established on the basis of the inseparability principle underlying the second example of body and soul. First of all, the soul and the physical world are considered as two distinct entities with permanent reality assigned to them. This is in total contrast to the position taken by Śaṁkara in his *māyā* doctrine in which he assigned only illusory existence to the soul and the physical world. Secondly, the principle of inseparability is applied to the three entities *Brahman*, soul and the physical world, recognizing that while the latter two coexist as two eternally independent entities, they are at the same time dependent on the eternal reality of *Brahman*. This is achieved by considering a new composite entity of soul and the physical world as the body of God, where God is used synonymously with *Brahman*. In this new body and soul conception, God is the independent reality controlling His body which consists of the individual soul and the physical world. The latter two are governed and sustained by God, and they exist entirely for the purpose of serving Him.

Rāmānuja's conception of the Absolute is very clear from the body and soul relationship. It is an *organic unity* such as we witness in a living organism where the control function is conspicuously present. There is a controlling element, and the rest of the elements are subservient to it. The dependent elements are called *viśeṣaṇas*, and the controlling element is called *viśeṣya*. God is the *viśeṣya*, and the individual soul and the physical world are the *viśeṣaṇas* in our problem under discussion. The *viśeṣaṇas* owe their existence only within the complex whole (*viśiṣṭa*) which includes God as the independent entity. It is the complex whole that is regarded as a unity and is given the name *viśiṣṭādvaita*.

The *viśiṣṭādvaita* principle is encountered in numerous situations wherever we can identify the inseparability phenomena either between two substances or between a substance and its attribute. It is not merely confined to the discussion of reality as a whole which is, of course, our problem of immediate interest. For instance, in the earlier example of a 'rose is red', we recognize that the material substance of the rose consists of the two distinct entities of rose-ness and redness, and according to the *viśiṣṭādvaita* principle, the red rose is a unity. Another example of the principle is that of a person who was once young and who has now become old. In this case, we recognize that the soul embodying the person at the two different instants of time is the same, thus emphasizing the organic unity where the dependent entities, though distinct and inseparable, do not coexist but are displaced in time.

The word substance has a much broader meaning in Rāmānuja's principle of inseparability. It is understood as something that undergoes change or that exhibits several modes (*avasthāvat*). This list includes six kinds, the most important of which are *Prakṛti*, *Jīva* and God which enter into the *viśiṣṭādvaita* theme of conceptualizing reality.

The concept of *Prakṛti* in Rāmānuja's philosophy is very similar to that encountered in our discussion of Sāṅkhya philosophy, except for some minor differences. Consequently, we shall only comment on the latter aspects. First, matter is not regarded as independent of the spirit, and in fact, it is inseparable from it. This becomes a point of difference because, according to the conventional presentation of Sāṅkhya, *puruṣa* is depicted as an indifferent spectator of *Prakṛti*. Secondly, unlike in Sāṅkhya, the three *guṇas sattva, rajas* and *tamas* are considered as attributes of *Prakṛti* rather than its constituents. This change is necessitated by virtue of the principle of inseparability between substance and attribute which is totally alien to the Sāṅkhya doctrine. Also, God is considered the cause of the diversity of the physical world, a philosophy which again is a direct consequence of the *viśiṣṭādvaita* doctrine. The relation between *Prakṛti* and its evolutes is also governed by the inseparability principle. Rāmānuja calls this relation *satkāryavāda*, a term which emphasizes the fact that the primary existence, *sat*, is itself regarded as the effect due to the modal transformations of the evolutes. The transformation is always between modes so that the material cause itself is not a substance but a mode. Except for the aforementioned differences, the concept of *Prakṛti* is identical to its depiction in the Sāṅkhya doctrine. Again, it is interesting to note the pervasive influence of Sāṅkhya on other systems of philosophy.

In passing, it is interesting to note that the protagonists of the Sāṅkhya philosophy consider *satkāryavāda* as the common basis for both the Sāṅkhya doctrine and for Sage Patanjali's Yoga philosophy. In the case of Sāṅkhya, it forms the theoretical basis, whereas for Yoga it forms the practical basis. *Prakṛti* and *Puruṣa* are considered as the two basic principles arising from *satkāryavāda*. Rāmānuja's body–soul (*śarīra* and *śarīri*) relationship is, according to them, one of several analogies, though perhaps the most picturesque one, which explain the relationship that exists between *Prakṛti* and *Puruṣa*. Other such analogies used are: *Jada* (inert material) and *Cetanā* (consciousness), *ātmā* (Self) and *anātmā* (not-Self), *vināśi* (destructible) and *avināśi* (indestructible). In all these analogies, the qualities describing *Prakṛti* are subordinate to those of *Puruṣa*, which is the the primary concept.

While discussing Śaṁkara's *advaita*, we drew attention to the fact that one of the principal objectives of his commentary on *Vedānta Sūtras* was to refute the legitimacy of the absolutistic version of *Vedānta* called *Brahma pariṇāmavāda*. The criticism was based on the inapplicability of the law of causality between the transcendental realm of *Brahman* and the universe where space and time have meaning. Since Rāmānuja also has established a direct relation between *Brahman* and the two distinct realities of the finite self and the physical world, his law of causation has to be understood properly if it is to escape the kind of criticism that was leveled against *Brahma pariṇāmavāda* by Śaṁkara. The subtlety inherent in his interpretation of *satkāryavāda* has to be understood in this context. Here causality (*kāraṇatva*) means transformation of the causal substance into a new form and not into a new substance. Accordingly, the causality relation here is not between *Brahman* as the causal substance and the world as the effect, as in *Brahma pariṇāmavāda*. Rather, it is a relation between *Brahman* in the state of cause and *Brahman* in the state of effect. It is only a modal transformation of the same substance and hence *satkāryavāda* steers clear of the inherent contradictions of *Brahma pariṇāmavāda*.

We have already said that the *viśiṣṭādvaita* doctrine considers the soul as being eternal and distinct from God. The soul is called a *prakāra* of God a nomenclature which brings forth the idea that it is not a mode but rather something that is collateral but not identical with Him. Since, according to the doctrine, God is embodied in *Jīva*, His immanence is what is responsible for internal control and guidance. There is a plurality of souls, all sentient in nature, and their common characteristics are due to the fact that they belong to the same class. *Jīva* is innately happy because of the immanence of God, but it is subject to the *karma* doctrine with all its implications of strict moral retribution and transmigration. Three types of souls are recognized. First is the category where there is no experience of bondage at all, as in the case of *Garuda* which is the eagle that symbolizes total freedom at all times. Secondly, there is the category of those who have gone through the cycle of birth and death and have already attained freedom. The last is the category of those who are still in the long process of transmigration who constitute the target audience for all philosophical speculations and religious practices.

In the *viśiṣṭādvaita* doctrine, God is visualized as the cosmic soul, and the two distinct entities of *Jīva* and *Prakṛti* are regarded as his body. The three together constitute an organic unity. There is not only inseparability between Him and the two other entities, but His immanence ensures that He is the one that guides and sustains them. This phenomenon which is operating at the cosmic level is also operative at the level of the individual body and soul. For Rāmānuja, God was not an undifferentiated reality as it was for Śaṁkara. Differentiation was an absolute necessity in order to ensure perception by His devotee. We include below some quotations from John B. Carman's scholarly book on *The Theology of Ramānuja* [11] to convey Rāmānuja's concept of God. What follows is a definition of *Brahman* as it appears in Rāmānuja's commentary on the Brahmasutras, *Śrībhāṣya*.

> By the word '*Brahman*' is denoted the Supreme Person [*Puruṣottama*], who is by inherent nature [*svabhāvatah*] free from all imperfections [*doṣa*] and possesses hosts of auspicious qualities [kalyāṇa-guṇa] which are countless and of matchless excellence [*anavadhidakatisaya*]. In all contexts the term '*Brahman*' is applied to whatever possesses the quality of greatness [*bṛhattva*], but its primary and most significant meaning is that Being whose greatness is of matchless excellence, both in His essential nature and in His other qualities. It is only the Lord of all [*Sarveśvara*] who is such a Being. Therefore the word '*Brahman*' is primarily used to signify Him.

In Śaṁkara's *advaita*, *Brahman* was without attributes (*nirguṇa*). Here, *Brahman* is equated with God and His auspicious qualities are extolled. Thus,

Brahman becomes qualified in this thesis.

We quote again from Carman who presents an English translation of a key passage from Rāmānuja's *Vedārtha-saṁgraha* which deals with the resolution of the seeming differences in the scriptural texts about the relation between God and the cosmos;

> The scriptural texts that deal with the immutability of *Brahman* have their most significant meaning [*mukhyārtha*] by the very denial of modification of His essential nature [*svarūpa*]. Those stating that He is attribute-less are also well established since they pertain to the negation of the defiling qualities of material nature. Those that deny plurality are well ensured by the affirmation that all entities, both spiritual and material, are the modes of *Brahman* by virtue of constituting His body, and that *Brahman* having everything as His modes exists as the sole reality, because He is the Self of all. The passages speaking of *Brahman* as different from the modes, as Master [*Pati*], as Lord [*Īśvara*], as the abode of all auspicious qualities, as the One whose desire are eternally realized and whose will is ever accomplished, etc., are justly retained by accepting just that. Statements that He is sheer knowledge and bliss are maintained because they express the defining property of the essential nature [*svarūpa–nirūpaka–dharma*] of the Supreme *Brahman*, who is different from all, the support of all, the cause of the origination, subsistence, and dissolution of all, is sheer knowledge in the form of bliss opposed to any impurity. Therefore, His essential nature, being self-illuminating, is also completely knowledge or consciousness [*jñānam-eva*] . The declarations of unity are well founded, since by virtue of the body–soul relationship, the identification of the two realities in coordinate predication [*samānādhikāraṇya*] is seen to be the most significant meaning of these texts.

Rāmānuja sings the glory of the Lord by enumerating several of His auspicious qualities (*kalyāṇa-guṇas*) which in no way conflict with the essential nature of *Brahman* that is described by the three qualities of existence, consciousness and bliss. Some of these auspicious qualities are categorized into two groups by His commentators. The rationale for the two groupings is provided by the two facets of the reality associated with God. The first is His supremacy (*paratva*) and inaccessibility because He resides in the transcendental realm. The second group of qualities refer to His easy accessibility (*saulabhya*) to His ardent devotees out of His own compassion and mercy through His many incarnations. In other words, *paratva* and *saulabhya* are the two polarities we associate with Him.

Based on this interpretation, the auspicious qualities that connote the first polarity of *paratva* are His knowledge, untiring strength, sovereignty, immutability, creative power and splendor, whose respective Sanskrit equivalents are *jñāna*, *bala*, *aiśvarya*, *vīrya*, *sakti* and *tejas*. The second group of auspicious qualities consist of compassion, condescension, forgiving love and generosity, which are translated as *kāruṇya*, *sauśilya*, *vātsalya* and *audārya*. The latter set of qualities are precisely those that are necessary to establish fellowship between the devotee and his personal God. Within the conceptual framework provided by the two polarities of *paratva* and *saulabhya* and its associated auspicious qualities, Rāmānuja also emphasizes that *God is not only the goal to be achieved but also He is the very means to achieve that goal.* This conclusion has immense significance in the practice of spirituality since it refers to the two-way relationship that exists between the means of achievement and the goal to be reached.

The concept of divine sport (*lilā*) is also unique to Rāmānuja's doctrine. In theistic doctrines, one of the central questions to be answered is why should God, even though He is perfectly capable of doing so, bother to create this universe at all when an infinity of choices are available to Him. Rāmānuja offers a solution to this riddle on the basis of the concept of divine sport (*lilā*). The term *lilāvibhūti*, which is of significance to the Śrīvaiṣṇava literature, means the manifestation of God's power and rule. Since God is without any desires of His own and is completely perfect (*paripūrna*), He could have no other design in mind except merely as a *lilā*. Furthermore, He is untouched by the gross inequities of life, and He cannot be held responsible for the pain and tragedy that exists within the realm of His creation. The created beings are all subject to the *karma* doctrine, and it should be their responsibility to liberate themselves from the cycle of birth and death. The latter also implies that God's creation takes into account the constraint imposed by the *karma* doctrine which is that each soul in a body is created commensurate with its past deeds. Since the *karma* doctrine is entirely deterministic in nature and acts as an apparent constraint on the divine sport, one can also conclude that the divine sport also is not aimless and haphazard but has a well-defined plan of its own.

The personal God is known as Nārāyaṇa which means the final resting place of all souls. His transcendental abode is *Vaikunta* in which He exhibits his eternal manifestation. His auspicious qualities ensure that because of His own innate nature of infinite mercy, He will take more steps towards His devotees than they can take towards Him through their unstinted devotion. He takes on various manifestations in order to redeem His devotees. One way which is special to the Śrīvaiṣṇava tradition is through *vyuha*, which is explained in the *pancarātra* literature. The second way of manifestation is through His *vaibhavas*, which are the well-known incarnations of Rāma, Kṛṣṇa etc. A third way of manifestation is by residing as the inner Self of all souls, that is, as their inner controller (*antaryāmi*). The fourth type of manifestation is implied in the concept of *arcāvatāra* in the consecrated images of the sacred temples.

The ideal of this theistic faith is the attainment of the abode of Nārāyaṇa to enjoy supreme bliss in His very presence. The means for realizing this lofty ideal are of two kinds, *prapatti* and *bhakti*. *Prapatti*, which is a special form of devotion is meant for all without any reference to the devotee's long and arduous preparation as specified by Vedic religions in general. It only asks for total surrender (*śaraṇāgati*) at the feet of the Lord with absolute faith in His qualities of mercy and compassion towards His devotee. As stated earlier, God is not only looked upon as the goal to be achieved but also as the means for attaining that very goal. But even to take this path, where the emphasis is on developing a certain orientation of the mind, one has to receive instructions from a qualified teacher since *prapatti* is also a form of knowledge (*jñāna*). While advocating *prapatti* as a means of salvation, Rāmānuja does not minimize the importance of the traditional path of devotion (*bhakti*) based on the *Upaniṣads*. Since this entitlement is within reach only for those who have gone through a prior discipline of moral and ethical cleansing, it is naturally restricted to a much smaller group than those who resort to *prapatti* as the chosen path.

The difference between *prapatti* and *bhakti* is sometimes explained on the basis of the duality that exists between divine Grace and self-effort. In *prapatti*, the emphasis is either exclusively on divine Grace, as in the *Thengalai* school, or on a minimum of self-effort coupled with a maximum of divine Grace, as in the *Vadagalai* school. These are the two sects within the Śrīvaiṣṇava faith. The small difference that exists between these two schools is picturesquely described by two similes. The first school, in which divine Grace is the sole consideration underlying *prapatti*, is compared to a cat carrying its offspring where the latter is at the exclusive mercy of the former. The minimum effort that is demanded of the second type of *prapatti* is compared to the meek submission of a baby monkey which is holding on to its mother with minimum effort.

In the case of *bhakti*, there is an emphasis on self–effort also, although the final release is ascribed to divine Grace only. The self-effort consists of three successive stages: *karma yoga*, *jñāna yoga* and *bhakti yoga*. One of the most extensively discussed subjects in Hindu philosophy is the importance assigned to the roles of action (*karma*), devotion (*bhakti*), and knowledge (*jñāna*) in the attainment of spiritual liberation (*mokṣa*) which is the proclaimed final aim of life. The Bhagavadgītā, which is perhaps the most popular holy book of the Hindus, presents a comprehensive treatment of the subject. Both Śaṁkarācārya and Rāmānujācārya have written commentaries on the Gītā following their philosophical insights of *advaita* and *viśiṣṭādvaita* and these works are considered of seminal importance to their doctrines. We shall present a brief summary of the philosophy first from the point of view of *advaita* and later point out the significant differences in Rāmānuja's interpretation.

In section 3.5, while discussing the four aims of life, we already dwelt on the need for religious ethics (*dharma*) for achieving both tangible and intangible benefits of our actions. A life led on the basis of religious ethics has the ability to provide the controls that are necessary for the pursuit of security (*artha*) and pleasures (*kāma*) during the natural processes of life, and they constitute the tangible benefits. Since man is bestowed with the freedom to act, unlike an animal which has to depend entirely on instinct, he needs to resort to some voluntary constraints that are well proven by time and tradition in order to exercise his freedom properly. Such actions based on religious ethics, in addition to the tangible benefits they confer, also have the certain promise of bestowing *pūṇya* (religious merits) which are the intangible benefits. This topic has already been discussed in connection with our exposition of the *karma* doctrine. But the important philosophical observation arising from our present discussion is that even a life based on the most assiduous practice of religious ethics does not in itself contain the potential for achieving spiritual liberation. This conclusion is arrived at by noting that the doer (*kartā*), who is performing the action, has a limited capacity by virtue of his finitude, and also by the recognition that the nature of an action, or in fact a series of actions, can only lead to limited results. The infinite consciousness, which is what spiritual liberation represents, cannot, consequently, be realized within the field of action which is necessarily finite.

The field of action, however, is extremely important for achieving refinement of the mind, and it is precisely a refined mind that can achieve the final goal of attaining spiritual salvation. In order to understand the mechanics of this process of refinement, it is necessary to be cognizant of the fact that it is our likes (*rāga*) and dislikes (*dveṣa*) that color our reactions, in thought and deed, to situations. Since it is not possible to sit in judgment over one's own likes and dislikes, let alone attempt to eradicate them directly, one has to devise a clever indirect strategy for accomplishing the goal in gradual steps. This is accomplished by realizing that there is a definite way for neutralizing the effects of likes and dislikes. The process of neutralization is dynamic in character and will of necessity take place when one is within the field of action. The explanation of this process takes us to some of the basic concepts of devotion.

In chapter 1, we concluded that the laws of nature constitute the element of constancy of the universal mind, which is the field of being at the cosmic level. We can restate the same conclusion in simpler terms making use of the language of devotion by recognizing that the laws of nature are nothing but the infallible laws of God. Being aware of this central truth, we can now make certain statements about what exactly is involved when a man acts with all his likes and dislikes intact. When a man acts with the freedom bestowed on him by virtue of his free will, he should know that the results of his action are nevertheless subject to the laws of nature, that is, to God's will. If the laws specific to an action are already known in advance, then man's actions will accordingly be based on this additional information. For instance, an action pertaining to walking on a mountain peak will be guided by the prior knowledge of the law of gravity which acts as a constraint to the arbitrary use of free will. In all such cases, the results of one's actions can be anticipated with a great measure of certainty.

However, there are many situations in life in which we act to fulfill our desires without any prior knowledge of the results of our actions because of our utter ignorance about the laws that govern such actions. From the perspective of the totality of man's actions, this represents the usual situation. The uncertainty of the consequences of making a particular career choice is an example of this category of actions. All that we know is that we have some specific result in mind for whose fulfillment we have performed the action. In addition we have the constant awareness that the results of our actions are indeed guided by the laws of the Lord. In other words, we have control only over the choices we make regarding our actions, whereas we have absolutely no control over the results of our actions. The outcome of the action might very well correspond to what we had expected in which case we feel elated; or alternatively, we get frustrated when there is no such correspondence, all because of our bundle of likes and dislikes.

The strategy for neutralizing such likes and dislikes is based on the simple recognition that since the results of our actions are not in our hands, whatever God wills should be accepted with perfect equanimity. We should accept the results as His Grace (*prasāda*). The conscious cultivation of this mental attitude towards actions we perform born out of deep devotion to God, will when it becomes an ingrained habit, have the pronounced effect of increasing the efficiency of our actions since the perceptual filter composed of our likes and dislikes will not act as an obstruction. What is ultimately called for is a change in mental attitude while performing actions. This change should result in the total surrender of the 'I' notion, which is what gives legitimacy for the doer-ship in action, to the will of God. This act of supreme devotion, when fulfilled, will automatically ensure the realization of the full potential of the field of action.

Surrender of the ego factor at the feet of the Almighty is not achieved merely by the intellectual recognition of its necessity for achieving salvation. Love and devotion to the Lord will certainly have the propensity to gather increased intensity with the march of time. There are, however, certain austerities that are recommended in order to strengthen the resolve of the seeker. These are suggested in addition to the discipline concerning the cultivation of the right attitude for indulging in the field of action. In the ascending order of importance, the three important disciplines are prayer and worship (*pūjā*), chanting of devotional hymns in praise of the Lord (*stotra*), and meditation (*dhyāna*). Meditation is the subtlest of the disciplines that is practiced for accomplishing refinement of the mind (*citta-suddhi*) and its one-pointedness (*ekāgratā*). We have already discussed this topic in chapter 3.

It is important to emphasize at this stage that despite the importance of meditation in spiritual practice, it has to be recognized that it is by no means the final goal. It belongs to the category of means albeit the most important one. This point is worth noting because of the ambiguity that surrounds its discussion in some quarters. After attaining transcendence (*samādhi*) as a result of meditation, the seeker has to return to his normal waking state of consciousness. More often than not, this return to the waking state is also signified by a return to spiritual ignorance. The bluntness with which this thought is expressed is not meant to minimize the role of meditation but rather to lay emphasis on its proper role.

In the ultimate analysis, spiritual ignorance can only be overcome by its contrary spiritual knowledge. Fulfilling the ideals of action and devotion will assist in the inquiry about the fundamental question of ' who am I?' This inquiry has to be assisted by the declarations of the scriptures, which serve as an independent testimony (*pramāṇa*), and the guidance of a preceptor, in order to avoid the inherent difficulty of the mind inquiring about its own true nature. Such a constant inquiry, with a knowledge of all its ramifications, will ultimately result in unleashing a mode of the mind which will dispel all spiritual ignorance once and for all even when one is in the conscious state of mind. In popular discourses, this is referred to as the intuitive faculty of the mind in contrast to its rational faculty. In philosophical literature, the experience of the intuitive mind is called the plenary experience (*akhandākāravṛtti*) and is considered to be the final fruit of the field of knowledge. Spiritual knowledge thus acquired will be a permanent accomplishment and constitutes the realization of the supreme truth in one's own life time.

The three achievements are in the fields of action, devotion and knowledge, called *karma yoga*, *bhakti yoga* and *jñāna yoga*, respectively. The word *yoga*, which appears in all the three fields, refers to the union of the soul (*jīva*) with the supreme consciousness, as in the non-dualistic interpretation. An equivalent interpretation in the devotional literature is in terms of the intimate bonding that exists between the seeker and the Creator. The question of union (*yoga*), however, arises only when there is first a clear experience of the duality that exists between the real 'I' and the other three states of consciousness of the seeker. It is important to note that this transcendental duality is very different from the duality expressed in the Descartes statement *I think, therefore I am.* The latter duality between mind and matter can easily be comprehended on an intellectual basis, and as discussed in chapter 4, forms the starting point of philosophical investigations concerning man and nature. As for the experience of transcendental duality, it is meditation that gives one the clear recognition of this phenomenon, and so meditation plays a key role in all the discussions concerning the achievement of the final goal in life.

Finally, in this brief discussion about the role of action, devotion, and knowledge in the realization of the final aim of life, we observe that there is no fundamental difference between supreme devotion (*bhakti*), which asks for a surrender of the 'I' notion, and spiritual knowledge (*jñāna*) which identifies the Self with the supreme consciousness. In the former case, there is the overweening conviction that everything in the universe is subject to God's will. Surrender of the ' I' notion is also born out of the realization that even free will that is experienced, is in the ultimate analysis, governed by the laws of the Lord. In the second case of spiritual knowledge, the emphasis is on the inquiry about the Self and the final realization that there is no difference between the Self and *Brahman.* The role of action (*karma*) is complementary to the aspects of both devotion and knowledge. We also note that the pursuit of *mokṣa* calls for the transcendence of both Ethics and Logic since the pursuit of religious ethics, though necessary, is not sufficient for the purpose. Similarly even the limits of rationality cannot generate the intuitive mode of the mind necessary for the ultimate spiritual experience. We have already noted in chapter 6 that in Śaṁkara's *advaita*, *mokṣa* is considered attainable in one's own life, which is an optimistic conclusion emphasized by the concept of *jīvanmukta.*

We shall now present the principal differences in Rāmānuja's doctrine from those of *advaita.* Here, we find that the sequence is in terms of *karma yoga*, *jñāna yoga* and *bhakti yoga*, which means that the aspect of devotion comes as the final stage of culmination. *Jñāna yoga*, which comes next to *karma yoga*, has a special significance in the *viśiṣṭādvaita* doctrine. The purpose of this yoga is two-fold. First, it is to realize the true relation of the self in relation to God, and second, the relation of self with *Prakṛti* which is God's cosmic vesture. The means for achieving this self-realization is through meditation and also by the realization that the self is entirely subordinate to God as prescribed by the *apṛthak–siddhi* principle. The difference in emphasis between *advaita* and *viśiṣṭādvaita* comes to the fore in this interpretation of *jñāna yoga*. In *advaita*, self-realization is by itself the ultimate goal, whereas in this doctrine, it is only a precondition to God realization.

Bhakti Yoga is the culmination of the three stage process. The word *bhakti* means devotion born out of love for God without expectation of any rewards. It is achieved as a result of the highest spiritual knowledge gained by a practice of the earlier stages. It asks for loving meditation on Him, but unlike other systems of philosophy, it does not promise a perception of the ultimate reality. What is promised, on the other hand, is a 'beatific vision of the Divine' as a result of the intense love showered on the object of meditation. The ultimate goal is realized only after discarding the physical body when the soul has a direct vision of God. Since this vision does not depend on the senses, it can be described as an unique non-sensory experience. Here again there is a difference from the *advaita* school where the latter recognizes *jīvanmukti*, which is realization when one is still alive. Many followers of the *viśiṣṭādvaita* faith view the culmination of the doctrine as an anticlimax since they would have much preferred the end goal of *jīvanmukti* as in the *advaita* doctrine.

Rāmānuja's theistic philosophy had a profound influence on the *bhakti* movement in general. For instance, Swami Rāmānanda (1300–1411) in North India, who is considered to be the dominant figure for spreading the gospel of the *bhakti* doctrine, is believed to have been a *viśiṣṭādvaitin*. His twelve disciples, who were well-established spiritual leaders, spread the message all over the country. Also, the doctrines of Nimbārka, Vallabha and Caitanya have all the common kernel of the *viśiṣṭādvaita* principle embedded in them, since they are different shades of the latter doctrine that emphasizes identity while admitting differences.

We stated earlier that Rāmānuja placed great importance in integrating the main tenets of Vaiṣṇavism into his *Upaniṣadic* doctrine. These tenets are based on a system called *Pancarātra*, which also is deemed to derive its authority from the *Bhāgavata* system of the pre-historic period. The system is grouped under four headings: a section on knowledge (*jñāna pāda*), a section on meditation (*yoga pāda*), a section on installation of temple images (*kriyā pāda*), and finally, a section on ethics (*charyā pāda*). The ultimate purpose is to saturate the mind of the devotee so that all his acts become acts of worship dedicated to the Supreme God Nārāyaṇa. Rāmānuja's school of philosophy, which is also an *Upaniṣadic* school, does not recognize any conflict at all with the *Pancarātra* system.

In conclusion, we include some comments on the significance of Rāmānuja as a philosopher and theologian. This is considered important since there exists a widespread erroneous impression that Śaṁkara's *advaita* is the only authentic exposition of *Vedānta*. The majority of Hindus who are interested in acquiring some theoretical understanding about their philosophy and religion depend upon translations of the Sanskrit texts either in their regional languages or in English. Most of these translations have devoted the majority of their coverage to *advaitic* literature, and so there is a compelling reason why such a misunderstanding could occur. This is particularly so in the West since translations into western languages are almost exclusively devoted to the *advaita* philosophy. If Rāmānuja's philosophy ever gets mentioned, it is only to convey the limited impression that his principal objective was to refute Śaṁkara's philosophy. The fact of his being an eminent philosopher and theologian in his own right never receives its due credit.

Even some Indian philosophers who are also knowledgeable in Western philosophy have referred to Rāmānuja as a 'Hindu Theologian', thereby suggesting that he was primarily a theologian rather than a philosopher. The eminent philosopher S. Rādhākrishnan, who was once the president of India, has also emphasized the role of Rāmānuja as a theologian. The word theologian needs some clarification since it is used differently in the context of the great religions of the world. In the case of Rāmānuja, his description as a theologian does not mean that the *viśiṣṭādvaita* school is the result of a creed carved out of a dogmatic formulation of *Vedānta* totally bereft of reason. Rather, the reference is based on his success in incorporating the central tenets of the devotional faith of Vaiṣṇavism into his reading of the *Upaniṣadic* texts. This success does not by any means belittle his importance as a philosopher of the theistic faith of *viśiṣṭādvaita*. This is why his followers consider him not only as a great philosopher but also as a saintly figure who gave practical guidance to his religious community.

7.3 Madhvāchārya: Dualistic School

Madhva also belongs to the theistic school of Vedanta, but he preached the philosophy of *Dvaita* which means dualism. The important tenets of *Dvaita* are conspicuously different from those of the earlier schools we have discussed.

Madhva came later than Shankara and Ramānuja, and lived for 79 years from 1238 to 1317 AD. He was born in a Brahmin family in a village close to Udipi in the present day state of Karnataka. He was called Vāsudeva by his parents, but after he became a renunciant, he assumed the name of Purna Prajna, meaning, ' the fully enlightened'; a change of name is customary when one enters a spiritual order after renouncing the life of a householder. Till today, the *mutt* (monastery) that he had established for the propagation of his branch of Vedantic philosophy remains fully alive and vibrant. The deity that he had installed, the famous Udipi Krishna, continues to be an important place of worship which draws pilgrims from all over the country. The soul-stirring songs composed in the Kannada language by the saintly musician, Purandara Dasa, in his ecstatic devotion to Krishna are highly popular even today. Madhva was an ardent Vaishnavite (worshipper of Vishnu) , and his attitude towards Shaivisim (worship of Shiva) was uncompromisingly combative. He ridiculed the concept of *Māya* which is a central concept of the non-dualistic school, and referred to Shankara as a crypto-Buddhist to indicate his strong disapproval of *advaita* being classified within the orthodox school of Vedanta.

Madhva's spiritual preceptor was known to be a scholar in *Advaita* philosophy, but the student had serious differences of opinion with his master about the interpretations of the scriptural texts. But realizing Madhva's intellectual brilliance and high level of spiritual perfection, his teacher installed him as the head of his own mutt when he was only 16 years of age. One of the first tasks that the young head of the *mutt* undertook was to go on a missionary tour with the purpose of engaging Jains, Buddhists and Advaitins in debates in order to win them over to his dualistic interpretation of Vedanta. The ruling monarch of the day was also supposed to have participated in this missionary activity, not through debates, but through the exercise of kingly power.

Following the tradition of Shankara and Rāmānuja and other great teachers that preceded him, Madhva also has written commentaries on the three important texts of Brahma Sutra, Gīta and the Upanishads to establish his philosophy of dualism on a firm scholarly footing. Madhva was a prolific writer, and he wrote under the name of Ānandathīrtha, leaving behind more than thirty major works including commentaries on Bhāgavata Purana and portions of the great epic Mahabharata. Many teachers of Vedanta attach great prominence to Bhāgavata since this voluminous text makes a conscious effort to popularize the abstract spiritual message by rendering it into a more palatable form through the literary medium of mythological stories. Madhva's commentaries highlight the dualistic doctrine that supposedly underlies the purānic stories.

The main features of Madhva's dualistic philosophy can be put into a concise form based on our understanding of the earlier Vedantic theses of Shankara and Ramānuja , and the theses of of Sankhya and Yoga that precede them. We intend to give only an introductory idea of the dvaita doctrine, and examine its implications in the explanation of the concepts of a) soul (*jīva*); b) nature (*prakriti*); and c) God (*Iswara*) . Also of interest is how the above three concepts are interrelated. In this connection, we focus attention on the concept of *bhēda* (difference) which is a specialty of the doctrine. Further, we turn our attention to determine how the dualistic doctrine explains the famous Upanishadic statement *Thou art That* which is at the center of controversy between the various vedāntic commentaries. Other points of interest are whether the philosophy is idealistic or realistic, which law of causation it subscribes to, the number of *pramāṇas* that are admitted in the validation of knowledge, and the manner in which it explains error in perception; these are highly technical details of any Vedantic philosophy, but we will not dwell on them in our very brief exposition of the *dvaita* philosophy.

Being a theistic doctrine, its description of the relation of God, man , and nature, is made from the relativistic plane of existence. The diversity of nature is considered real, whether it pertains to individual souls or to nature in its infinite variety. It means that one soul differs from another soul, and also any two entities of the inanimate world differ from one another. Despite this infinite diversity, each and every soul, and each object of the inanimate world is considered unique. The manner in which it substantiates the uniqueness of every entity is based on the concept of *bhēda* (difference) which is a special feature of the doctrine. The validation of this special concept of difference to explain uniqueness is based on a rather involved logical argument, but suffice it to know that the *dvaita* doctrine states that uniqueness of a thing is a consequence of its difference from other things.

The doctrine, being theistic, places emphasis on a personal God which, in this case, is Vishnu, a deity worshipped from early vedic times. The belief in God is derived entirely from the authority of the scriptures, and therefore no appeal is made to the logical argument based on *bhēda* which is reserved exclusively to explain the uniqueness of the entities of Nature. It is important to note that the *pramāṇa* used for substantiating reality of the phenomenal world is never invoked to substantiate the truth of the existence of God. God is considered as the efficient cause for creation of this universe whereas the material cause is ascribed to *prakriti* (nature), which is in conformity with the duality that is characteristic of Sankhya philosophy (see chapter 5) where we considered *puruša* and *prakriti* (spirit and matter) as the dual pair.

The description of Nature as a whole is more or less similar to its conception in the Sankhya philosophy except for a small difference. In the case of Sankhya, the three guṇas of *sattva, rajas* and *tamas* were considered as attributes of prakriti. Here, in the *dvaita* philosophy, they are considered as the very first evolutes of nature. From then on, the manner in which the evolutes unfold is the same as in the Sankhya schematic (see Figure 5.1).

Returning to the concept of difference, the *dvaita* philosophy identifies five types of differences: the differences between individual souls, the differences between entities of the inanimate matter, difference between God and soul, difference between God and Nature, and, finally, difference between soul and matter. The reason for enumerating these differences is to provide the basis for the logical argument to establish the uniqueness of the three entities: the individual soul, an arbitrary entity of inanimate matter, and God. Once this uniqueness is established, the doctrine proceeds to point out that uniqueness does not imply the independence of these various entities. It is God, who is immanent in the relativistic plane of existence, that is considered to be totally independent of all the rest of the entities. It is He who controls everything in the universe from within.

God and revelation are intimately connected. The doctrine attributes all the exquisite divine opulence to its personal God. But it also points out that even when God reveals himself, the spiritual aspirant cannot exhaustively know all his auspicious qualities. This reminds us of the scene in Gīta where Lord Krishna reveals his cosmic personality to Arjuna but he complains that he is unable to see all His infinite features. There is a difference made between apprehension of the Lord and his total comprehension by the devotee. God is not only the creator and destroyer, but also a controller of all aspects of the universe. The diversity of nature is only a revelation of his perfection. Nature, in all its diversity, affords opportunities for spiritual aspirants to take concrete steps to attain spiritual perfection, but, in the final analysis, God's grace is absolutely necessary for individual efforts to fructify. In consonant with the concept of divine grace, a key idea of theism, the doctrine also believes in the concept of *avatar* which renders it possible for the humans to come into contact with the divine; more specifically, in *dvaita* philosophy, an *avatār* is the mediator between Vishnu and the humankind. Also, the doctrine admits of the possibility of God taking different *avatārs* to fulfill different purposes in his creation at different times in the evolution of the universe.

As for the concept of *jīva* (soul), even though its uniqueness as an entity of nature is established, it has, nevertheless, certain common features with God.

From our earlier discussion, we recall that *jīva* consists of a spiritual element that is combined with the internal sense organ of an individual; the latter is made of the three subtlest elements of nature, which are: *buddhi* (intellect); *ahamkāra* (egoism); and *manas* (mind). The spiritual element, which is the element of constancy in *jīva*, possesses the features of sentience and bliss which are also the intrinsic features of God. According to the doctrine, these common features in God and soul are by no means identical, but they bear only a strong resemblance to each other. We recall that in Shankara's *Advaita*, the features of sentience and bliss in God and soul were construed as identical, an important point of difference from the *dvaita* conception which substitutes resemblance for identity. It is the concept of resemblance between *jīva* and God that provides the basis for explaining the famous Upanishadic statement *Thou art That* whose interpretation has given rise to so much controversy in the several Vedantic theses. In the case of the *dvaita* doctrine, Madhva explains the statement on the basis of *resemblance* between *Thou and That.*

As a slight digression, we wish to briefly comment on how the entities of inanimate matter are viewed in the doctrine. Although the uniqueness of any single entity of matter is well established on the basis of *bhēda*, however, it is generally assumed that it does not possess any spiritual element associated with it. But one can question this conclusion on the basis of our present day knowledge of particle physics where we see the particles also seem to flaunt a degree of independence in their movements subject only to probabilistic laws. It means that spirit coexists with every entity of nature, whether animate or inanimate. This observation might appear like a hair-splitting argument, but we submit that it is something to ponder over. But, even if we modify our conclusion in the light of what modern physics suggests, it will not vitiate the main trend of the *dvaita* thesis, where emphasis is laid on the absolute dependence of both soul and matter on God's will.

It is worthwhile to point out that the *dvaita* doctrine is completely realistic in the sense that it holds that there can be knowledge only with reference to an object. Anything that exists in time and space is considered real regardless of one's perception about it. This is in contrast to the absolutistic doctrine of non-dualism where there is a distinction made between different orders of reality, between the absolute and relative. It is this ontological difference that prompts *dvaita* to make light of the concept of *māya.*

As in earlier Vedantic theses, *avidya* (spiritual ignorance) is considered to be the reason for an individual for not being able to gain release from the chain of birth and death (*samsāra*). In *dvaita*, spiritual ignorance is considered to be real having a dual effect: it not only blocks the vision that is necessary for realizing the true nature of God, but it also obscures the true vision of oneself. The latter deficiency can be overcome by the study of scriptures, accompanied by the steadfast practice of meditation combined with a life of moral rectitude. The realization of God, on the other hand, can only be achieved through his infinite grace. Knowledge of oneself is considered as mediate knowledge and it can only become immediate only when one experiences God realization through his grace. The dualistic philosophy emphasizes the importance of *bhakti* (devotion to the Lord) as the final means for reaching him. Uninterrupted love and devotion to the Lord stems from an unfaltering faith in his compassion. The decisive cause for salvation is due to God's grace *prasāda* showered on the devotee who is constantly trying to reach him through his *bhakti.*

Part III

Science & Spirituality: A Discussion

Chapter 8

Spirituality: The Universal Message of Vedic Philosophy

Our purpose in discussing various metaphysical doctrines in the earlier chapters was to provide insight into some of the principal teachings of Indian philosophy which subscribe to the authority of the Vedic scriptures. These are also called the orthodox schools, and sometimes, for convenience, they are simply referred to as Hindu schools of philosophy. The Sāṅkhya–Yoga pair is grouped under the classification of orthodox schools, or the Vedic schools, whereas Śaṁkara's *advaita*, Rāmānuja's *viśiṣṭādvaita*, and Madhva's Dvaita are further classified under *Upaniṣadic* or *Vedāntic* schools. What is conspicuous from our discussions is that there is considerable commonality of final aims amongst them despite their sharp differences in conceptualization. The commonality is exhibited in both transcendence and the immanence of the undifferentiated consciousness, that is, of the Brahman and the Ātman of *Vedānta*, although this central idea is expressed differently in the Sāṅkhya–Yoga schools. Since direct instruction regarding the non-manifest reality is an impossibility, the only way left to gain spiritual knowledge is to resort to suitable methods of indirect instruction. The differences in the various metaphysical doctrines also serve the practical purpose of catering to different levels of preparation (*adhikārabheda*) in the seekers.

Each metaphysical doctrine that we have discussed is complete in itself as to its final purport and therefore has the intrinsic appeal for a special group of people. Despite this broad understanding, the followers of each doctrine find it necessary to refute the others based on the conviction that their own doctrine is meant for the most learned of the seekers who are after spiritual perfection. So we are left with the option of making up our own minds as to our preference for a doctrine even though we attest to the truth of the Vedas about the ultimate realities of *Brahman* and *Ātman*. We have no intention of engaging in the futile task of endorsing the superiority of one doctrine over another although our earlier chapters are heavily influenced by the non-dualistic interpretation.

All the schools of philosophy we have discussed have their own ardent supporters some of whom speak from the lofty plane of their own personal experiences. We cannot conclude, however, that only one school could be the right one to follow because the differences amongst them only pertain to the final stages of the spiritual journey, that is, to the manner in which the *jīva* and the physical universe are related to *Brahman*. Moreover, these doctrines were put forward at different periods of time, and so there is always the question of whether some of the differences could not have been resolved if the chief protagonists had been contemporaries. For instance, in Śaṁkara's time, the Buddhistic influence was very much prevalent, and therefore he had to make a special effort to refute the Buddhistic concept of the void (*śūnya*). Rāmānuja placed a very high premium on the integration of the Vaiṣṇava traditions into his *viśiṣṭādvaita* doctrine, which posed some special problems of its own. Christianity had a viable presence in India by the time Madhva preached his Vedantic doctrine. The origin of Sāṅkhya philosophy can be traced to pre-Buddhistic times, and although its proximity to the *Vedāntic* schools cannot be denied, it suffers from the handicap of having lost some of the original texts which could have shed more light on some of the controversies.

The debates carried on by the followers of the *Vedāntic* schools, in the generations that followed, unfortunately carry the imprint of sectarian differences, a divisiveness which was never the intention of those doctrines in the first place. As stated earlier, the appeal that a doctrine holds for an individual depends on his own special aptitude for philosophical speculation and commitment to spiritual practice. For example, people with scientific training who are used to dealing with abstractions seem to have very little difficulty in appreciating the concept of *māyā* of *advaita* philosophy whereas it may be abhorrent to someone whose natural proclivity is for a purely theistic interpretation. In the final analysis, the purpose of a doctrine is to assist one in overcoming one's own unique set of conditions of obscuration from spiritual knowledge. Once progress is achieved in this direction, the adherence to the particular doctrine of one's choice assumes secondary importance. Consequently, once the general picture is comprehended, what matters most is spiritual practice because only by experience can both intellectual clarity and the requisite faith be strengthened. The conceptual framework of one's choice can, at best, provide an effective means for indulging in constant contemplation about the ultimate truths that can, in turn, reinforce confidence in both the theory and practice of spiritual life. On the other hand, the study of the doctrines merely for satisfying intellectual curiosity will only have a very limited value and may, incidentally, serve the futile purpose of engaging in endless sterile debates of a sectarian nature.

In this concluding chapter, we shall indulge in a free-wheeling discussion of many topics allied to our main theme. In the process of broadening the scope of our discussion, we shall first present the main differences in approach between Western philosophy and Indian philosophy and also the place of science and religion in a general context. An understanding of science is important since all the doctrines invariably deal with some broad features of our physical universe, and in that regard, it is important that we always defer to the views of science whenever there is a point of difference on well–proven facts. For instance, there is no point in arguing for the religious view that there were always a fixed number of species on this planet when the theory of evolution is very clear about the contrary position. A metaphysical doctrine should not depend for its validity on such patently absurd aberrations from scientific truth. Of course, science will not be of much help, at this stage at least, if the comment on the number of species is not confined to this planet alone. We do not yet know whether life exists elsewhere in the universe. The more important reason for bringing science into the picture is the impact of scientific paradigms on philosophical thought, which was the subject matter of chapter 1. As for religion, we have not included its detailed discussion in our plan, except to point out how it relates to science, philosophy and spirituality.

The origins of Indian philosophy go back almost thirty centuries, if not more. Unfortunately, the very poor historical record of this period has given rise to serious gaps in our understanding at very critical places despite intensive research. Also, as we have seen from our earlier discussions, accurate biographical data of even some of the towering philosophers is, lamentably, practically non-existent. We do not know for sure, for instance, the exact dates of Sage Kapila, whose Sāṅkhya philosophy has had such a pervasive influence on several other doctrines, but only that it is very ancient, perhaps of the pre-Buddhistic period. The only date that we know for certain during the first millennium of the history of Indian philosophy is the date of Lord Buddha's death in 487 B.C. Undoubtedly, the discussion of Indian philosophy would have a greater impact if we had access to reliable historical records. This situation stands in marked contrast to the history of Western philosophy, which prides itself on the accuracy of its records.

Indian philosophy refers to all the schools of philosophy originating in India.

Consequently, apart from the schools associated with Hinduism, it also includes philosophies of the ancient national religion of Jainism and the world religion of Buddhism. But our study was restricted only to the schools of Hindu philosophy, which come under the Vedic fold. Our plan was simply to sample the principal ideas of some of the major schools within this class in order to provide different conceptual frameworks for the exposition of the central theme of Vedic philosophy. To do so we dealt with the ancient doctrine of Sāṅkhya due to Sage Kapila, the Yoga philosophy of Sage Patanjali, the Absolutistic school of *Vedānta* due to Śaṁkarācārya, and the Theistic schools of Vedanta due to Rāmānuja and Madhva. Our purpose was also to weave in some key ideas from the history of science to fortify the treatment of ancient philosophy. We believe that the new imagery that we have projected will have an appeal to the modern mind.

There is a close kinship between philosophy and religion because both are concerned with probing the question of the central meaning of existence. The obscurity in the meaning of the word religion arises because it admits a wide variety of interpretations in actual practice. In the name of religion, one could cultivate a belief system that is based purely on theological dogma, or it is also possible to entertain a burning desire for the realization of God. The outward observances can also vary enormously, from, for example, a serene meditation hall to a busy temple where there is an endless jostling of devotees given to blind stirrings of devotion. Typical examples of these extremes are the places of worship of the *Vedāntic* Societies of the Rāmakṛiṣṇa Mission where the atmosphere is always tranquil and very conducive to silent contemplation, and, in marked contrast, the extremely busy ambience of the Tirupati temple in Andhra Pradesh, where, during any time of the day or night, there will always be a sea of humanity patiently waiting for hours on end to have a fleeting visit (*darśana*) with their patron deity (*iṣtadevatā*). It is a moving experience to listen to those prayers replete with inexpressible groaning.

Science is also a close companion of philosophy and religion although its overlapping domain of inquiry is different. Science deals with the exploration of truths underlying our external universe, and although it does not pretend to unveil the ultimate mystery surrounding creation, it does attempt to answer through comprehensive theories all the questions posed since that event. In chapter 1 we discussed several of the important theories of physics, cosmology, and biology which provided a number of paradigms with a direct bearing on the study of the *ultimates*. Scientific endeavors in this regard are the following: in cosmology, the goal is to examine questions associated with the origin and evolution of the universe; and in biology, the goal is to discover the origin and evolution of life, and if possible, to determine the causes of human consciousness and self-awareness. The enterprise of science extends in the direction of both unity and diversity; on the one hand, there is a constant effort to discover unifying laws to explain diverse phenomena, and on the other hand, the ever diversifying universe affords abundant opportunities to discover new phenomena. For example, the Theory of Relativity and Quantum Mechanics belong to the first category with their emphasis on unity, whereas the splitting of the atom by Rutherford belongs to the second category.

We can begin to appreciate the dimension of universality in the endeavors of science and spirituality only when we do not restrict ourselves either to the understanding of our external world or to our own internal universe of thoughts based exclusively on what our common sense can support. If we did, we could not have gauged the significance of the scientific theories that we discussed in chapter 1 or appreciated the need for metaphysical doctrines. This need to look beyond one's immediate environment in search of either an intellectual understanding or a deep personal experience is expressed in various ways by scientists and philosophers. In Professor Whitehead's words, there is a "connection between universality and solitariness" inasmuch as "universality is a disconnection from immediate surroundings". Such a disconnection is particularly conspicuous in the case of religion, whose feature of universality is combined with the intense desire on the part of the individual for a personal experience almost impervious to the environment.

Of late, incredible as it might seem, several conferences have been held in India where prominent theologians and scientists from around the world have participated with the sole purpose of exploring the possibility of synthesizing the underlying ideas of their specialties. In all these discussions, the scope of religion has included spirituality, which is a common factor in all the principal schools of philosophy arising from India. It is this inclusion that provided a central core of ideas for purposes of comparison with the goals of science. We have seen that, in spite of the differences in the metaphysical conception of the schools that we have discussed, the goal of attaining a higher state of consciousness has been common. Despite the lack of success in achieving a synthesis, which did not come as a surprise to any of the experts in the various disciplines, these conferences have rendered yeoman service in bringing to the fore several points of congruence between science, religion and spirituality and have indirectly promoted a healthy respect for both. The viewpoint expressed by Fritjof Capra in his *Tao of Physics* [10] best summarizes the outcome of such deliberations:

> I see science and mysticism as two complementary manifestations of the human mind; of its rational and intuitive faculties. The modern physicist experiences the world through an extreme specialization

> of the rational mind; the mystic through an extreme specialization of the intuitive mind. The two approaches are entirely different and involve far more than a certain view of the physical world. However, they are complementary as we have learned to say in physics. Neither is comprehended by the other, nor can either of them be reduced to the other, but both of them are necessary, supplementing one another for a fuller understanding of the world.... Science does not need mysticism and mysticism does not need science; but man needs both.

In Sanskrit, as already stated, there are two descriptive words to connote the two distinctive domains of secular and spiritual knowledge. Secular knowledge is called *aparā vidyā* (lower knowledge) and scriptural knowledge is called *parā vidyā* (higher knowledge). The words lower and higher do not signify any kind of water-tight compartments since the entire field of knowledge is considered to be one continuum. The terminologies are meant to emphasize that knowledge about the non-manifest field of existence, the mystical experience, is considered higher in comparison to the relative knowledge that deals with the realities of our external world. While individual goals are invariably different as far as the pursuit of secular knowledge is concerned, the spiritual goal of all mankind, on the other hand, is considered to be one and the same, namely the realization of the Self, which is the innermost essence of man. This was the central thrust of all the philosophical doctrines that we have discussed.

The valid means for acquiring secular knowledge as in science are of a separate category altogether and stand in contrast to the means that are employed for acquiring spiritual knowledge. According to this broad division of knowledge, all secular knowledge, which includes scientific knowledge in particular, is categorized as *aparā vidyā*, whereas the aim of religion assigns it the status of *parā vidyā*. This two-way division of knowledge in the philosophical literature of India clearly points out the impossibility of achieving a synthesis between science and mysticism. This division does not, however, mean that knowledge of science is irrelevant for the understanding of religions based on mysticism. In fact, the Vedic religion, has always highlighted the paramount importance of analytical thinking, rationality, and experimentation in developing sharper insights into intuitive knowledge because they have the rich potential to serve as useful pointers to our non-manifest field of existence. The experiences of mystics of all religions throughout history inform us that the amphibious nature of the human mind is certainly capable of crisscrossing the domains of knowledge through both intuition and intellect. In fact, the optimistic conclusion is that there will be a heightened awareness of the external world in all its myriad aspects when the mind develops the ingrained habit of frequenting the non-manifest field of existence through meditation and other allied practices.

There is no inherent contradiction in simultaneously embracing both religion and science, provided, of course, that the scope of the two subjects and their line of demarcation is properly recognized. If adherence to religion is entirely due to theological dogma without any experience of its principal pronouncements, there would, of course, be plenty of room left for a serious internal conflict to arise between the values of science and religion. This is likely to be the case in a religion where mysticism has lost all its hold. On the other hand, if the understanding of religion is rooted in a metaphysical system espousing the existence of a transcendental reality whose veracity is vouchsafed by men and women of proven integrity and whose declarations also contain an assurance of the possibility for verification by those who are determined to tread the spiritual path, then it cannot, in good faith, be subject to the blanket criticisms that are routinely leveled against received wisdom and theological dogma. Even though the process of verification within the spiritual realm is of an entirely different kind than in scientific studies, it is not likely to be lightly brushed aside even by those devout scientists who do not like anything that even remotely smacks of blind faith. Fritjof Capra has written extensively in his *Tao of Physics* about the exploration of scientific and spiritual truths and has succeeded in casting serious doubts about the dismissive argument of subjectivism with which scientists often charge the latter. In fact, there is now a new group of well-known scientists both in the West and in the East who have taken keen interest in keeping up with the exploration of the nebulous area between science and mysticism, proceeding from their side of the interface area.

Even if one were to acknowledge the irreconcilability in goals between science and religion despite their common feature of universality, the difficulty is not carried over to the practitioners of science and religion. It is well-known that throughout the history of science, there have been eminent scientists who were also deeply religious, and conversely, many notable men of religion who did not blindly accept theological dogma. To cite only one example, Issac Newton, one of the greatest scientists of all times, was a deeply religious man and wrote prolifically about religion, although this contribution remained anonymous till his death because of the tenor of the times that he lived in. He kept his extensive treatise on religious subjects secret for fear of being mistaken as anti-scientific. As for men of religion being possessed by an uncompromising faith in their scientific discoveries, we need only recall the heroic story of Galileo who doggedly held on to his convictions grounded in empirical observations of the outer reaches of space. He did so despite the trials and tribulations he had to face from an authoritarian Church.

In the category of men of religion having regard for scientific methods, one has only to take cognizance of the rich traditions of Indian philosophical thought. All the six systems of philosophy in the orthodox category, that is, those that subscribe to the authority of the Vedas, underscore the importance of rational thinking, which is characteristic of science. For instance, one of the complaints that we normally hear about Śaṁkara's philosophy of non-dualism is that it is so abstract and theoretical that it appeals only to intellectuals. As for the Yoga philosophy of Sage Patanjali, which is usually paired with the Sāṅkhya philosophy of Sage Kapila, it is entirely devoted to the practical means of experiencing the revealed truths. In scientific terminology, the Yoga philosophy belongs predominantly to the category of experimentation where the principal scientific apparatus used is the seeker's own psycho-physical apparatus. Although the concept of verification of hypotheses in the spiritual realm is of a different type altogether than that applicable in scientific experiments, the insistence on suspending judgment until one actually experiences the axiomatic statements of the theory is likely to invite the healthy respect of a scientist who is full of skepticism for mystic practices.

Next, we shall lead up to a brief discussion of the fundamental differences between Eastern and Western philosophy. In marked contrast to those of the West, the Indian schools of philosophy do not make any specific appeal to the history of natural sciences although they do accord a primacy of place to the role of scientific rationality in matters concerning worldly realities. In their metaphysical doctrines, however, there is virtually no reference to the natural sciences as we understand them today. It is this singular absence that was puzzling to the Western philosophers, so much so that they hesitated for a long time to include Indian philosophy as a legitimate field of study even in university departments of religion devoted to inter-religious studies. The reluctance was due mainly to ignorance about the nature of the differences rather than prejudice. Western philosophy traces its roots to ancient Greece, and, until lately, it was quite oblivious to the philosophical thinking of the East. This "spiritual iron curtain" that was drawn in ancient Greece caused the unfamiliarity in the West with religions rooted in India, an impediment which has begun to be overcome only within the last century. On the other hand, because of historic reasons, India has had access to knowledge from Judaism, Christianity, and Islam. However, Indian philosophical thought developed in total isolation of Western thought.

In the meantime, there is great interest in India to explain Vedic concepts in terms of the paradigms of modern day science for purpose of achieving better communication. One can never deny the need as time progresses for new imagery to convey old ideas. For example, metaphors borrowed from computer technology have spread to all disciplines, and the discipline of religion is no exception. When one says that all the religions of the world are like different software designed to achieve the same goal, the underlying thought is likely to be understood instantly by most people. Because of this attraction in new ways of communication, it is not uncommon to find Vedic scholars who exhibit a great deal of enthusiasm for using the paradigms of science to better communicate their ideas. In their enthusiasm, some have gone overboard by glossing over the subtle differences that exist between scientific and philosophical reasoning while transferring analogies from science to philosophy. Nevertheless, we can surely pardon knowledgeable Vedic scholars who flirt with modern science and thereby unknowingly mix up scientific reasoning with purely philosophical arguments in their exuberant attempt to enhance the appeal of their discourses to those who are steeped in a culture of science and technology.

On the other hand, what is really significant is that there are a number of Vedic scholars who have presented unmuddled expositions because of their equal familiarity with both science and Vedic religion. As a rule, those scholars who are also ardent practitioners of their faith and who have made progress in their spiritual practice present the philosophical ideas with great clarity. As opposed to these, the scholars who are merely dependent on their library knowledge without a scintilla of spiritual experience are readily obvious in their shallowness, notwithstanding their sincere efforts to bring in more clarity by appealing to some major discoveries in science. Authentic expositions of a religion based on mysticism seem to require spiritual practice as a necessary prerequisite since, in the ultimate analysis, what one is indulging in is an articulation of an ineffable experience through the unavoidable use of language and reasoning.

In the above context, it is important to note the fundamental differences that exist between the central concepts of Western and Eastern philosophy. Since our discussion pertains to the schools of Eastern philosophy associated with the Vedic religion, we shall, for purposes of comparison, continue to assume the key idea of mysticism, namely, that it is possible to transcend our ordinary states of consciousness in search of the core meaning of human existence. As should be evident by now, while Western philosophy is very much influenced by the natural sciences, Eastern philosophy is much closer to mysticism and is therefore hardly influenced by the natural sciences. The ultimate aim of *parā vidyā* is true spiritual knowledge, which is equated with religious experience. As such, metaphysical arguments appear as part and parcel of the religious doctrines. Indian philosophy has never been relegated to the realm of mere intellectual speculation on the mysteries surrounding man and the universe. In contrast to this view, we present an extended quotation from the mathematician– philosopher Bertrand Russell from his introduction to *A History of Western Philosophy* [39] to glean some understanding of the viewpoint of the West on the subject:

> Philosophy, as I shall understand the word, is something intermediate between theology and science. Like theology, it consists of speculations on matters as to which definite knowledge has, so far, been unascertainable; but like science, it appeals to human reason rather than to authority, whether that of tradition or that of revelation. All *definite* knowledge–so I should contend–belongs to science; all *dogma* to theology. But between theology and Science there is a No Man's Land, exposed to attack from both sides; this No Man's Land is philosophy. Almost all the questions of most interest to speculative minds are such as science cannot answer, and the confident answers of theologians no longer seem so convincing as they did in former centuries. Is the world divided into mind and matter, and, if so, what is mind and what is matter? Is mind subject to matter, or is it possessed of independent powers? Has the universe any unity or purpose? Is it evolving towards some goal? Are there really laws of Nature, or do we believe in them only because of our innate love of order? Is man what he seems to the astronomer, a tiny lump of impure carbon and water impotently crawling on a small and unimportant planet? Or is he what he appears to Hamlet? Is he perhaps both at once? Is there a way of living that is noble and another that is base, or are all ways of living merely futile? If there is a way of living that is noble, in what does it consist, and how shall we achieve it? Must the good be eternal in order to deserve to be valued, or is it worth seeking even if the universe is inexorably moving towards death? Is there such a thing as wisdom, or is what seems such merely the ultimate refinement of folly? To such questions no answer can be found in the laboratory. Theologies have professed to give answers, all too definite; but their very definiteness causes modern minds to view them with suspicion. The studying of these questions, if not the answering of them, is the business of philosophy.

Having defined the scope of Western Philosophy, one is naturally inclined to question why anyone would be interested in pursuing such abstract inquiries into what, for all intents and purposes, seem to be insoluble problems. Russell provides two reasons in support of philosophical studies: one is dependent on a historical perspective of human civilization from time immemorial and the other on his purely personal motivation that is not atypical of all thinking individuals. As for the historical reason, there is irrefutable evidence that man's actions throughout history have been guided by the then prevailing theories on life and the universe at large, and also on the ethical concepts of what constitutes good and evil. In fact, a particular historical period of a nation cannot be clearly understood without a general understanding of its current philosophical outlook, and reciprocally, only those who possess a philosophical faculty can ever hope to gain a proper understanding of it. Since philosophy has been identified as the No Man's Land between science and religion, one can expect significant changes in philosophical outlook to occur with changes in the paradigms of science.

For instance, there were bound to be differences in philosophical perspectives before and after the times of Galileo, Newton, Darwin , Einstein, and Planck. After Copernicus, the ego- and geocentric view that man was at the center of the universe had to be abandoned. After Newton, scientists in various disciplines looked for strictly deterministic ways of explaining all phenomena, maintaining a clear separation between the observer and the observed; the workings of the universe came to be viewed in a purely mechanistic way as analogous to that of a gigantic clock. Darwin's theory of evolution gave such a rude jolt to the Biblical ideas of evolution and genesis that the controversy between evolutionists and creationists has remained unresolved. Einstein's concept of space-time put an end to a basic assumption of the German philosopher Immanuel Kant who advanced the idea in *The Critique of Pure Reason* that space and time represented two separate categories of thought constituting two *a priori* features. Because of Planck and other great quantum mechanicians, we now know that chance and randomness are intrinsic to the subtler layers of matter, thus ending the era of the universal appeal of deterministic theories to explain all natural phenomena. Closely associated with this idea is a fact of immense philosophical significance suggested by Heisenberg's uncertainty principle, which puts an end to the clear separation between the observer and the observed in the investigation of natural phenomena. Rather than considering man to be a disinterested observer, the theory demands that he cannot be viewed in isolation from the combined observer–observed relationship. We have dealt with these and other topics in some detail earlier in our development in order to bring home the idea that scientific paradigms have had a profound effect on the development of Western philosophy.

We will now draw attention to the one change that has occurred to philosophical thinking due to changes in the tenets of theology. We noted that mysticism was very much an integral part of early Christianity, but it lost ground in the later centuries because the voice of those who wanted to resurrect this feature was lost in the wilderness. But the presence or absence of mysticism in theology makes a world of difference in philosophical thinking. One has only to refer to that great classic in the religious literature of Christianity, *The Cloud of Unknowing* [48], written anonymously by a medieval British mystic. To quote Aldous Huxley from his book *Grey Eminence* [25]:

> 'The cloud of unknowing' is the same as what the Areopagite call the 'super- luminous darkness'–the impenetrable mystery of God's otherness. Ultimate reality is incommensurable with our own illusoriness and imperfection; therefore, it cannot be understood by means of intellectual operations; for intellectual operations depend upon language, and our vocabulary and syntax were evolved for the purpose of dealing precisely with that imperfection and illusoriness, with which God is incommensurable. Ultimate reality cannot be understood except intuitively, through an act of the will and the affections.

The essential message of the The cloud of unknowing is very similar to the theistic doctrines associated with Vedic philosophy although we do not wish to minimize the differences in Christian and Vedic theology. One can further verify the common theme underlying the message of the mystics by referring to the sayings of other great Semitic mystics. However, Pope John Paul II in *Crossing The Threshold of Hope* [32], has pointed out what he considers to be the distinguishing feature of Christian mysticism from the mysticism of the East.

> At various times, attempts to link this method with the Christian mystics have been made–whether it is with those from northern Europe (Eckhart, Tauler, Suso, Ruysbroeck) or the later Spanish mystics (Saint Teresa of Avila, Saint John of the Cross). But when Saint John of the Cross, in the *Ascent of Mount Carmel* and in the *Dark Night of the Soul*, speaks of the need for purification, for detachment from the world of senses, he does not conceive of that detachment as an end in itself. " To arrive at what now you do not enjoy, you must go where you do not enjoy. To reach what you do not know, you must go where you do not know. To come into possession of what you do not have, you must go where now you have nothing. In Eastern Asia these classic texts of Saint John of the Cross have been, at times, interpreted as a confirmation of Eastern Ascetic methods. But this Doctor of the Church does not merely propose detachment from the world. He poses detachment from

> the world in order to unite oneself to that which is outside of the world–by this I do not mean nirvana, but a personal God. Union with Him comes about not only through purification, but through love.

Pope Paul's remark about the differences between Christian mysticism and the mystic philosophies of the East arising out of the role of 'detachment' seems erroneous since Vedic philosophy also asserts that life cannot be led on the basis of detachment. Our discussion of the metaphysical doctrines should have cleared the misunderstanding from such an unfavorable characterization. More will be said about this later. For the present, even if we concede for the sake of argument that there are significant differences between Western and Eastern mysticism, what is central to this exposition is the importance a religion accords to mysticism. One can safely conclude that it is the continuity or discontinuity of mystic theology that has greatly contributed to the divergence of views reflected in the modern theological interpretations of these religions. When the important insight arising from mysticism is lost, there is always a tendency to cast the role of science and religion in terms of reason and dogma respectively. This rift has enhanced the appeal for philosophy to be much closer to the natural sciences than to religion. In contrast, mysticism has always remained the main well-spring of the philosophical doctrines of the Vedic religion right from inception, a fact which illustrates the pronounced differences between Western and Eastern philosophy.

The second reason that Russell advances for emphasizing the importance of philosophy is shared by many thinking individuals even outside the fold of science. The pursuit of any single branch of knowledge, which is what is within the reach of most people, instils a sense of humbleness because of the realization that what we actually know is only a tiny fragment of the total knowledge that is available to mankind. If this conclusion is true of Russell, who is reputed, from all accounts, to have possessed an encyclopedic mind, it is more true of lesser mortals. The chief characteristic of secular knowledge is its high degree of specialization. It branches out into disciplines and sub-disciplines *ad infinitum.* Since diversity of nature is a natural phenomenon, there are bound to be more and more specialization as time progresses. This thought is best codified in Dyson's principle of maximum diversity that we discussed in chapter 1. The scientist will never go out of business on this score, nor, in fact, will people of other disciplines. Scientific speculations about the possibility of arriving at a "Theory of Everything" to explain the totality of physical phenomena through a unified model, or about " The End of Physics" based on the conviction that the discipline had achieved its goal, shall forever remain figments of the imagination of overzealous scientists. New disciplines that the previous generation would not have even dreamt of will always appear on the horizon. The tremendous impact of digital computers and information technology on almost all current disciplines stands as testimony to this observation. The immense diversity of secular knowledge makes us aware of the wide gap that exists between what we actually know and the totality of knowledge available. This, in turn, triggers a heightened sensitivity to important things which lie outside the ambit of our own personal knowledge. On the other hand, a blind faith in a theology which makes definite declarations about the nature of the universe leaves no room for uncertainty and does not provide an intellectual climate where one can raise questions to resolve one's doubts. According to Russell, this dogmatic attitude, when in fact, we are groping in ignorance, is bound to generate indignation towards the unknown.

From the point of view of Vedic philosophy, which is much closer to religion than to natural science, Russell's criticism of theology appears unduly harsh since in Vedic philosophy there is plenty of emphasis on the need to contemplate the declarations of its doctrines. The concepts of *śravaṇa*, *manana*, and *nididhyāsana*, which we discussed in chapter 3, exist precisely because we do not have to accept metaphysical truths on the basis of dogma. Furthermore, the very pedagogical style of the *Upaniṣadic* teaching, which is devoted to the culmination of all Vedic knowledge, is couched in terms of questions and answers between a sagely teacher and an exuberant disciple, and makes it abundantly clear right from the start that the teacher is not asking the student to blindly follow. In fact, the *Upaniṣadic* teacher revels in posing philosophical riddles that have the effect of jolting the consciousness of the student, and he does not even volunteer to teach until, and unless, the student formulates an incisive question. Thus, there is a complete openness about the continuing dialogue between the teacher and his disciple once the teacher is convinced that the disciple has become a serious seeker. Philosophy, from the Hindu perspective, has the final aim of establishing the unity of all knowledge by locating its main source either in one's personal God, as in the theistic doctrines, or equivalently, in plenary consciousness, as in the absolutistic doctrines where the emphasis is on reaching the highest level of consciousness. Importance is given to both the effort on the part of the seeker and also the aspect of divine Grace for the realization of the ultimate truth. Since theory and practice go hand in hand, one enriching the other, they are given equal importance, and consequently, there is no chance for blind faith to rear its ugly head during the process of spiritual transformation.

The similarities and differences between Western and Indian philosophy are summarized by Heinrich Zimmer [49] as follows:

Indian, like Occidental, philosophy imparts information concerning the measurable structure and powers of the psyche, analyses man's intellectual faculties and the operations of his mind, evaluates various theories of human understanding, establishes the methods and laws of logic, classifies the senses, and studies the processes by which experiences are apprehended and assimilated, interpreted and comprehended. Hindu philosophers, like those of the West, pronounce on ethical values and moral standards. They study also the visible traits of phenomenal existence, criticizing the data of external experience and drawing deductions with respect to the supporting principles. India, that is to say, has had, and still has, its own disciplines of psychology, ethics, physics, and metaphysical theory. But the primary concern–in striking contrast to the interests of the modern philosophers of the West–has always been, not information, but transformation: a radical changing of man's nature and, therewith, a renovation of his understanding both of the outer world and of his own existence; a transformation as complete as possible, such as will amount when successful to a total conversion or rebirth.

The principal concern of Indian philosophy, therefore, is to bring about a steady transformation in the individual seeker by making use of his inner resources so as to guide him towards the realization of supreme truth. Spiritual knowledge is considered something that is already immanent in an individual and only waiting to be unmasked by the termination of spiritual ignorance, unlike secular knowledge which has to be freshly sought. This is the basic difference between the two types of knowledge that accounts for the differences in methodologies for acquiring them. Accordingly, the aim of spiritual instruction is to present both the theoretical and practical aspects about the science of the ultimate truth, known as mysticism, and the efforts are directed towards unveiling the self-luminous knowledge that is ever present. The theoretical aspect of mysticism comes through the declarations of the Vedas (*śruti*), that is to say, through revealed knowledge, and also from the cohesive voices of sages throughout history, whose actual experiences (*smṛti*) lend credence to the declarations of revealed knowledge. That is why both *śruti* and *smṛti* receive equal reverence. For instance, Bhagavadgītā, the most popular text and a real gem amongst the scriptures comes under the category of *smṛti.*

The universal and everlasting message about the ultimate truth associated with the external world and the individual self is addressed to all mankind in general without regard to individual religious preferences. The practical aspects of *sanātana dharma* or eternal religion are best learnt from an authentic teacher who has tread the spiritual path, and, consequently, there is a great deal of importance attached to the Indian tradition of a *guru* (one who can remove one's spiritual ignorance) to serve as a reliable guide. Unfortunately, the true meaning of the word *guru* has been corrupted beyond recognition in its modern day usage in India. An old idea from the scriptures has been borrowed and trivialized. Every small music or dance school, for instance, has its own resident *guru*, who pitifully attempts to maintain some outward trappings of the ancient tradition appropriate to his or her own profession, which, nevertheless, amounts to a caricature of its original meaning. In the West, the word *guru* is used to refer to an expert whether he be a computer wizard or an acclaimed sex pervert. It seems as though one cannot escape from the mindless onslaught on rare spiritual concepts engendered by either an attitude of total irreverence or ignorance.

Returning to our earlier comparison of religious ideas with the paradigms of science, we wish to reiterate that mysticism is not an exclusive feature of Hinduism; it was also integral to all the great religions of the world. For those who subscribe to the philosophical school of non-dualism, some of the sayings in other religions also are very much reminiscent of their own conception of the non-manifest reality. D.S. Subbaramaiya [44] quotes the following teachings of Christianity and of Sufism of Islam. From Sufi teaching, "Grasp well the subtle fact; thou art that which thou seekest.... The foremost duty of the seeker lies in eliminating his own separated existence.... Thou-ness and I-ness pertain to our world. They do not exist in the region of the Beloved.... He is the only reality; futile is the assertion of any existence but His." In Christianity, the Ten Commandments, the beatitudes, and the statements of Jesus Christ such as " He who has seen Me has seen the Father," "I and my Father are one", " I am the way, the Truth and the Life", and " I am the Light of the world" are reminiscent of the implied *advaitic* current. Such observations from the great religions of the world led Aldous Huxley to write his famous book *Perennial Philosophy* [23] which made an enormous impression on philosophically-inclined Indians when it was first published. The experiences and utterances of the mystics of Buddhism, Judaism, Christianity, and Islam that are well documented in history have a great deal in common with those of the Hindu mystics. To quote Evelyn Underhill [46]:

> Almost any religious system which fosters unearthly love is potentially a nursery for mystics: and Christianity, Islam, Brahmanism, and Buddhism each receives its sublime interpretation at their hands. Thus St.Theresa interprets her ecstatic apprehension of the Godhead in strictly Catholic terms, and St.John of the Cross contrives to harmonize his in-

> tense transcendentalism with incarnational and sacramental Christianity. Thus Boheme believed to the last that his explorations of eternity were consistent with the teaching of the Lutheran Church. The Sufis were good Mohammedans, Philo and the Kabbalists were orthodox Jews. Plotinus even adapted–though with difficulty!–the relics of paganism to his doctrine of the Real.

It is this common theme of mysticism, sometimes referred to as mystical theology, that provides the common basis fundamental for comparison with the tenets of science. Capra's *The Tao of Physics* is typical of this line of inquiry. The value of the book stems from the fact that the author is both a high-energy physicist and also some one who has, according to his own admission, actually had the rare experience of glimpsing the majestic sweep of the transcendental state of consciousness; it is not a dry intellectual inquiry made possible by ransacking a library on world religions.

From this vantage point, one derives the conviction that ecumenical debates within the various denominations of a single religion or a broader understanding between different religions become meaningful only when spirituality is accorded its true place. In the absence of this centerpiece of the cosmic jig-saw puzzle, inter-religious debates, however well–meaning, are bound to have an air of superficiality about them. It is well documented that the monks of various religions can communicate well with each other without harboring any animosity or mutual recrimination, but sadly, such is not the case with their priests who are ever eager to cross swords with each other. Despite the recorded history of mysticism in all major religions, unfortunately, as of today, it is only Buddhism, Hinduism, and the Sufi sect of Islam that have maintained their emphasis on mysticism without a break. And in the case of Hinduism, as indeed of all religions indigenous to India,it is singular in the sense that the theory and practice of spirituality constitutes the main ideal of the religion. This, in turn, has resonated with all other religions that have tried to maintain the practice as their core belief. This is particularly true for Buddhism. Though Buddhism does not subscribe to the authority of the Vedas and has consequently been termed as a heterodox religion in the classification of religions arising from India, and though Buddhism does not even explicitly invoke the concept of God, thus subjecting itself to being termed atheistic, it has remained a kindred religion of Hinduism for historic reasons and because of the common importance accorded by both to mysticism. Lord Buddha himself is held in such high esteem by the Hindus that he is regarded as an incarnation of God. As for Sufi mysticism, some notable religious historians of India have speculated with a great deal of nostalgia that if only this form of Islam had prevailed in India, the clash of the two cultures associated with Hinduism and Islam which resulted in the partition of the subcontinent in 1947 could have been avoided.

Next, we turn our attention to some aspects of religious practice connected with the Vedic religion, all of whose metaphysical doctrines are rooted in mysticism. The wide gap that exists between the precepts and practices of the various religions is a matter of common knowledge; however, the problem appears to assume formidable proportions in the case of religions based on mysticism. Part of the problem arises from the enormous difficulty in broadening the scope of communication so as to reach the largest number of people. While their sages have spoken from the platform of a higher level of consciousness, the listeners have no alternative but to receive their messages while stuck in their ordinary waking states of consciousness without ever having a clue what transcendentalism means. This barrier in communication between the two planes of consciousness has invariably resulted in the disciples' misunderstanding of the correct meaning of the message, particularly after a long lapse of time. Because of this inherent difficulty in preserving the clarity of the message, the *Upaniṣadic* teachers invariably took great pains to limit the attendance of their discourses to those disciples who through long and arduous practices had earned their entitlement to receive the instruction. The prerequisites imposed on gaining *Upaniṣadic* knowledge were quite unrelenting, and, consequently, it came to be regarded as a well-preserved secret (*rahasya*) meant only for a chosen few. It is interesting to contrast this tradition with what Pope John Paul II [32] has to say about spreading the Christian Gospel: " Evangelization is also the entire *wide-ranging commitment to reflect on revealed truth.*"

Whatever legitimacy there may have been for preserving a Vedic tradition of preserving a high degree of selectivity for the propagation of the revealed truth, the spirit of the modern times runs counter to the perpetuation of such an elitist practice. As a consequence, *Upaniṣadic* knowledge is now not only meant for the exclusive few who wish to undergo a spiritual transformation as a result of a steady cultivation of the accompanying religious practice, but it is also available to any one who wishes to have some information about it. In this respect, it is treated as any branch of secular knowledge, although the motivation for its study is very different. Because of this change in perspective, Indian philosophy now has the same wider audience in mind as Western philosophy has.

There are two broad aspects for achieving the goals prescribed by *Upaniṣadic* knowledge. The first is the intellectual understanding of the transcendental reality concerning man and the universe; the second is the essential aspects concerning moral and ethical cleansing. Two types of doctrines associated with *Upaniṣadic* knowledge can be identified. First there is the kind of doctrine which keeps the highest level of consciousness as the ultimate goal of religious experience; these are called absolutistic doctrines. Śaṁkara's non-dualism, which we discussed in chapter 6, was of this variety. Secondly, there are those doctrines which emphasize the notions of a personal God, a seeker's loving devotion for Him and the aspect of divine Grace which are characteristic of theism. We discussed Rāmānuja's *viśiṣṭādvaita* doctrine and Madhva's *dvaita* doctrine as typical of this category. The two sets of doctrines are closely related; in the first variety, the aspect of spiritual knowledge gets greater emphasis, whereas in the second, the aspect of devotion is the predominant feature.

The distinction that is usually made between devotion and knowledge to indicate separate paths for realizing the ultimate truth is incorrect. Supreme devotion lays stress on the blissful nature of *ātman*, whereas knowledge emphasizes the destruction of spiritual ignorance. Evelyn Underhill's [46] picturesque words describe such differences between the two types; in the first we are talking about the "marriage of the soul" whereas in the second we are talking about the "betrothal of the soul." For all practical purposes, it becomes a difference without a distinction. Whether one subscribes to the absolutistic or theistic doctrines, one of their common features is meditation, which is the effective means of transcending ordinary states of consciousness. Meditation is deemed as the chief instrument for overcoming spiritual ignorance. In addition, moral and ethical cleansing are also considered to be very essential accompaniments to spiritual practice. Interestingly, it has now become commonplace to find people who attempt to seek spiritual experience by concentrating exclusively on the aspect of meditation while paying scant attention to the complementary aspects of religious practice that are necessary for cleansing the "doors of perception." Teachers in India, however, point out that the lifestyle of such seekers is inadequately disciplined to achieve the goals they have set for themselves. While these traditional teachers are prepared to accede to some necessary adjustments in practice because of the times we live in and the cultures we belong to, they do not, however, go to the extent of giving their nod of approval either for the total dilution or abject abandonment of time-honored practices.

It is a matter of common observation that very few amongst the many who seriously take up spiritual life ever get to the stage of perfection that is theoretically possible to achieve. To quote Aldous Huxley [23]:

> But, like all other studies and practices, those of mystical theology must begin at the beginning. And the beginning is a long–drawn pro-

> cess of moral amendment, discursive meditation and training of the will. Hence the paucity of mystics; for the world is mainly peopled by Micawbers, optimistically convinced that something or somebody will turn up and get them out of the difficulties from which, as a matter of cold fact, they can be saved only by their own efforts. Many, in this case all, are called; but few are chosen, for the good reason that few choose themselves.

Earlier, after presenting Bertrand Russell's view on the scope of the subject, we touched upon the answers to the question of why anyone should be interested in studying Western philosophy at all. We can now ask ourselves a similar question with regard to Vedic religion based on spiritual practice. If it is true that only a very few of the practitioners can ever hope to attain the goal of spiritual enlightenment, the question can legitimately be raised as to why anyone should even bother to commence the spiritual journey. Implicit in this line of questioning is the assumption that success is to be measured in terms of one's ability to reach the goal here and now and anything that falls short of it should be deemed a failure. There is, of course, the traditional answer to this troubling question, which is to assure the seeker that the efforts made in his lifetime will not be wasted even if he does not reach the journey's end, because the merits accrued in this lifetime will place him in a more favorable position for resuming the spiritual journey after his next birth. The solace derived from the vague promise of a heavenly reward for earthly virtues appears to non-believers as a comforting fairy tale, but that is what is offered by the *karma doctrine*, without which so many ticklish philosophical questions remain unanswered. And we know there is no way of verifying the assertions of this doctrine with the methods of investigation that we normally employ in scientific pursuits; we must appeal to the Vedic testimony. Because of this intangibility, we wish to suggest the more pragmatic view that even the interim benefits of spiritual practice in the present life are quite attractive. In addition, the veracity of this claim has the possibility of having verified during one's own life span.

Once one commences the spiritual journey with recourse to the daily experience of the higher state of consciousness, however fleeting it may be, one develops the habit of keeping up with the pleasant practice for its own sake without consciously looking for the distant goal about which we only have assurance on the authority of the Vedas and the common declarations of mystics all over the world. The rejuvenating process itself lends, most assuredly, ever-increasing charm to life in general and, consequently, provides a great deal of internal contentment. When the sun rises in the morning, one looks forward eagerly to the meditative process, and the tranquillity experienced invariably has the effect of spilling over to the rest of the day. It is almost like recharging one's life-battery on a daily basis by connecting it to an inexhaustible source within. As time progresses and as the practice matures along with other aids to experience, whatever intellectual doubts that creep in from time to time will gradually melt away of their own accord, and the proverbial meandering mind will propel itself, more often than not, to a state of complete stillness and total tranquillity. All this internal transformation is likely to take place with a minimum of effort. It is as if the mind were returning to its natural state of deep rest without the least necessity of any internal guidance or external control, very much like the homing instinct of birds.

Mystics all over the world have spoken lyrically about such experiences, and if one could experience even a tiny portion of it, as indeed one can, it would make the spiritual process very worthwhile. Based on this point of view, we can state axiomatically that the *spiritual process itself serves as the interim goal.* When one starts enjoying the process for its own sake, one is not unduly worried about whether or not one attains the final goal that is stipulated on the basis of scriptural knowledge. Deep within, however, a conviction sets in that the process of increasing intensity in which one is indulging may indeed have a point of convergence.

We shall now briefly touch upon a second reason for embarking on the spiritual process. Hindus have often been savagely criticized for upholding the ideal of a life of negation because of the perception that they accord more importance in their religious practices to the inward alchemy of the mind than to the development of the mental faculties needed for leading a purposeful life in the affairs of the world. This perception has made the religion notorious for being ' detached', ' other worldly', or 'fatalistic'. Though such criticisms represent a grotesque distortion of the truth, it is helpful to understand why such mistaken impressions have arisen at all. Of course, life cannot be lived on the basis of total negation or helpless resignation. Towards the end of the nineteenth century and the beginning of the twentieth century, Swami Vivekānanda went all over the world with missionary zeal to allay all such misconceptions by pointing out the robust optimism that is characteristic of the Vedic religion. He did it as only he could, with his matchless eloquence and indomitable courage. But with the lapse of time, the necessity has arisen to repeat his message. It will not suffice to simply note that this truth has already been told and retold in the scriptural texts, and hence it is up to the critics to find out the truth. The fact of the matter is that it cannot be denied that a skewed understanding exists not only outside but within India. In view of clarifying this position, we shall make the next few comments on the relationship between secular and spiritual life.

The real purpose of spiritual life is to ensure a fully integrated life by means of which one could participate in our day to day world with enormous zest and unfailing optimism. In fact, when one is firmly connected to one's bottomless source of spiritual energy within, it is possible to conduct oneself in affairs of the world with greater force than would otherwise be possible. This is because the three fields of thinking, speaking, and acting, which are responsible for leading a purposeful life, continually get refined through the safe reliance on spiritual life. These fields are ultimately anchored to the Being, and even fleeting experiences of it have the profound effect of increasing their efficiency in performance. We hear about courses being conducted on 'positive thinking' as if thinking is the ultimate basis for speaking and acting. This technique might very well have some positive influence on participants because of thinking's order of precedence amongst the three fields, but effortless changes of significant magnitude can occur when it is realized that the ultimate basis is the Being. Accordingly, it is that imperishable source that has to be roused first.

In the English commentaries on Indian sacred texts, words such as detachment, renunciation, and duty are quite common, and they are, of course, proper in the contexts in which they appear. They are qualities which naturally sprout as a result of leading a disciplined spiritual life; on the other hand, they are not meant to serve as modes of outward behavior for leading a spiritual life. For instance, dispassion (*vairāgya*) arises as a natural result of spiritual practice; an artificial mimicking of this type of behavior can never serve as a means for successful spiritual practice. That would be nothing short of rank hypocrisy. When such virtues are wrongly practiced without any reference to spiritual life, they will, of course, take on the stamp of negativism. When philosophical concepts are turned upside down through mistaken understanding, they, unfortunately, carry negative overtones, and a person who is not exposed to Indian culture is bound to make incorrect inferences. Consequently, there is a need for greater amplification of the ideas about means and ends in order to minimize the chances of misrepresentation.

One other factor that may have contributed to the negative impressions is the severity with which some exponents of the Vedic tradition condemn the transitory nature of our phenomenal existence. The criticism is oftentimes so blunt and heavy-handed that it spawns a disgust for life. The real purpose would be better served by gently diverting one's attention to aspects of existence which have a more permanent nature. Again, when the context in which the criticisms occur about our worldly existence is forgotten, they give the unintended impression that they are aimed at denying a healthy secular life. We stated earlier that both *parā vidyā* and *aparā vidyā* are important because they form one continuum. But somewhere in the process of diverting attention to *parā vidyā*, which is the knowledge of the non-manifest field of existence, we might have unwittingly diminished the importance that we attach to knowledge of the worldly realities. The robustness of the Vedic doctrines and the vitality of the Hindu religion can come into full focus only when such impressions are eradicated from people's minds.

Closely associated with the above general observation is the role of science and technology vis-à-vis the pursuit of spiritual knowledge. Any expositor of Vedic knowledge who disdains the benefits of science and technology has to be viewed with suspicion as to his authenticity. Some people have the habit of heaping criticism and ridicule on these important endeavors because they firmly associate these fields with the wanton neglect of basic professional values that are necessary for safeguarding the survival of our human civilization. The litany of complaints is very long and includes the possibilities of nuclear disaster, ruthless use of weapons of mass destruction, misuses of genetic engineering, the mindless perpetration of voracious consumerism that bodes ill to all mankind, the rapid pace at which fossil fuels are being exhausted, the thinning of the ozone layer of our planet with all its horrible implications, and the repeated promise that there is a technological fix for all our serious problems. Yet, at the same time, these critics make no excuses for utilizing all the benefits of the scientific age in their own day-to-day lives. One cannot but help being scornful of such hypocrisy particularly when such irresponsible criticisms come from men of religion.

While it is important to criticize some of the harmful trends of science and technology, it is not correct to place the entire blame at the doors of scientists and technologists. Wisdom would dictate that we separate the ideals of science and technology from some of the pitfalls in their applied fields in the same way that we separate the ideals of religion from some of its baneful practices. All around us, we see armed conflicts in the name of religion between groups of people and nation states, and we are even forewarned that after the demise of communism, the real threat to world peace is religious fundamentalism. All these gloomy predictions, however, have not led the majority of the world's population to question the value of religion. When the first space shuttle of the erstwhile Soviet Union returned safely to earth, Nikita Krushchev made the sarcastic remark that his country's cosmonauts had not come across any angels in the sky, which was his way of making light of religious beliefs. But immediately upon the cessation of the cold war, we started to witness Russians returning in droves to their churches, mosques, and synagogues. Even seventy years of communist rule in a state declared atheist had not succeeded in extinguishing the innate urge of the Soviet people to return to their religious lives once they breathed the slightest whiff of political freedom. The problem, therefore, is not to minimize the importance of either religion or science, but to devise ways of avoiding the harmful deeds done in their names. In any case, the general public has only a vague understanding of science and the metaphysical basis for religion. The importance of science is understood mainly through the impact of technology on society, and the value of religion is felt only because of what has been passed on from generation to generation within a family and not by the received wisdom of one's particular faith.

It is not uncommon to come across people who declare themselves to be atheists for a variety of reasons. Their lack of belief may be due to their deep convictions in science, which almost worships the absence of finality on matters of great moment, or to their detestation of theological doctrines which do not leave room for uncertainty. It is also possible that they arrived at their conclusion because of unspeakable atrocities committed in the name of religion. They may also find the bigotry involved in championing the superiority of one religion over another loathsome. There are many honest and decent people in the world who believe that the gulf between the precepts and practices of religions cannot be bridged, and therefore, they see religion as a divisive force in our society which is already torn by many other divisions. Interestingly, both atheists and believers have the same proportion of people with high moral and ethical standards in their ranks, and so the believers do not have an edge over atheists on this score. This observation should lead one to the conclusion that it is not necessary to believe in God and religion in order to lead a sound ethical life. In fact, as we have stated earlier, Vedic philosophy upholds that it is necessary to transcend both logic and ethics in order to pursue spiritual life. The point that is made is that logic and ethics can serve only as means to an end and not an end in themselves. In fact, according to Vedic philosophy, the true basis of ethics (*dharma*) itself lies in the truth of the transcendental reality.

It is interesting to survey the opinions of atheists in India, some of whom have written forcefully about these issues in scholarly publications. These criticisms arise almost exclusively from social rather than metaphysical reasons. In the absence of an organized church for Hinduism, there has never been any reason for these critics of God and religion to impose voluntary constraints on themselves for fear of invoking the wrath of the orthodoxy. Usually, these criticisms are leveled against the gross abuse by the general public of the notion of a personal God, which, in most cases, is invoked only for petitionary prayers. Or, they stem from deep compassion for people who despite their devotion to God suffer economic hardship. That He should remain untouched by all the suffering in the world is something unimaginable to atheists. It is hard to argue about metaphysical matters with somebody who dissolves into a welter of tenderness from even the slightest contact with tragedies around him, particularly when one can identify with the reasons for such acute sorrow. But this is precisely what Vedic philosophy is all about. It is cognizant of the reasons for sorrow and suffering in this world and points a way out. The question of God's existence begins to make sense only to those who are deeply involved in probing the purpose of human existence.

This premise is more dramatically expressed in Buddhism. Only one saying attributed to Buddha is enough to make the point; he is reported to have said, ' All the waters of all the seas are not to be compared with the flood of tears which has flowed since the universe first was'. The teaching does not remain pessimistic because his four noble truths and the eight fold path are meant to show an escape door for human misery. In fact, as a starting point for the discussion, being aware of the anguish of life is common to all the *Vedāntic* schools. But this is not the main burden of the teaching since it shows a way to escape the suffering and to lead a very optimistic life. The philosophical assertion has always been that it is only by putting an end to *avidyā* that one can ever expect to achieve an enduring solution to the chronic problem that is afflicting mankind.

The *Upaniṣadic* God that we have dealt with in our earlier discussion of the various metaphysical doctrines is very different from the popular notion of God, which is the target of criticism of these well-meaning atheists. As for religion, it is said that it could mean anything ranging from " a sum of scruples which impede the free use of our faculties" [20] to a burning desire for a union with God. Certainly, one would be loathe to subscribe to the former type of religion. When we have a clearer understanding of the views of atheists, it is quite possible even for an ardent believer, to agree with them in some respects. In any case, believers in the Vedic religion and its ideals are not expected to entertain an evangelical fervor to convert others to their viewpoint. This is one of the reasons why Hinduism has been acknowledged as a very tolerant religion. This is perhaps the best heritage that we could maintain in spite of the tremendous odds that the Hindu society faces because of mounting sociopolitical pressures to depart from its rich tradition.

We shall conclude this chapter with a quotation from Professor Hiriyanna's book on *Outlines Of Indian Philosophy* [21]:

> The *Vedānta* may accordingly be taken to represent the consummation of Indian thought, and in it we may truly look for the highest type of the Indian ideal. On the theoretical side, it stands for the triumph of Absolutism and Theism, for whatever differences may characterize the various *Vedāntic* schools, they are classifiable under these two heads....

On the practical side, the triumph of *Vedānta* has meant the triumph of the positive ideal of life. This is shown not only by the social basis of the ethical discipline which the *Vedānta* as an orthodox doctrine commends, but also by its conception of highest good which consists, not in isolating the self from its environment as it does for the heterodox schools but in overcoming the opposition between the two by identifying the interests of the self with those of the whole. Both ideals alike involve the cultivation of complete detachment; but the detachment in the case of *Vedānta* is of a higher and finer type. Kālīdāsa, who, as the greatest of the Indian poets, may be expected to have given the truest expression to the ideal of practical life known to Indians, describes it as ' owning the whole world while disowning oneself'. The *Vedāntic* idea of the highest good also implies the recognition of a cosmic purpose, whether that purpose be conceived as ordained by God or as inherent in the nature of Reality itself, towards whose fulfillment everything consciously or unconsciously moves.

References

1. Barrow, John D; and Tippler, Frank J., The Anthropic Cosmological Principle, Clarendon Press, Oxford, 1986

2. Bawra, Brahmarishi Viśvātmā., Essence of the Gītā, Divine Radiance Publications, Pinjore, Haryana, India, Translation from Hindi to English by Rajendra Dubey, University of Waterloo, Canada

3. Benson, Herbert., The Relaxation Response, Avon Books, 1976, NY

4. Bridgman, Percy Williams., The Way Things Are, Harvard University Press, Cambridge, 1959

5. Campbell, Joseph., The Masks of God: Oriental Mythology, Penguin Books, 1962

6. Chardin, Teilhard., The Phenomenon of Man: with an introduction by Julian Huxley, Translated by Bernard Wall, Harper, NY, 1959

7. Dayānanda Swami., Talks On Upadeśa Sāram, Ramaṇa Maharishi, Translation from Tamil to English

8. Campbell, Joseph with Bill Moyers., Doubleday, 1988

9. Bohm, David., Wholeness and The Implicate Order, Ark Paperbacks, London, 1980

10. Capra, Fritjof., The Tao Of Physics, Collins: London, 1975

11. Carman, John B., The Theology Of Rāmānuja, Yale University Press, 1974

12. Chinmayānanda, Swami., The Science Of Life, Parts 1 and 2, Central Chinmaya Trust, Bombay, 1979

13. Chinmaya Trust., Shankara The Missionary, Bombay, 1978

14. Davies, Paul., God and The New Physics, Penguin Books, 1983

15. Davies, Paul., The Cosmic Blueprint, Simon and Schuster, 1989

16. Davies, Paul and John Gribbin., The Matter Myth, Simon and Schuster, 1992

17. Davies, Paul., The Mind of God, Penguin Books, 1992

18. Dyson, Freeman., Infinite in all Directions, Harper and Row, 1988

19. Hawking, Stephen W., Brief History of Time, Bantam Books, 1988

20. Hiriyanna, M., The Essentials of Indian Philosophy, Allen and Unwin, 1949

21. Hiriyanna, M., Outlines of Indian Philosophy, M. Hiriyanna, Allen and Unwin, 1932

22. Hofstadter, Douglas., Gödel, Escher, Bach: an eternal go, Basic Books, 1979

23. Huxley, Aldous., Perennial Philosophy, Collins: London, 1972

24. Huxley, Aldous., Ends and Means, Chatto and Windus: London, 1957

25. Huxley, Aldous., Grey Eminence, Meridian Books: New York, 1959

26. Mahesh, Yogi Maharishi, The Bhavad-Gītā, Penguin books, 1969

27. Osborne, Arthur., The Teachings of Bhagavān Śrī Ramaṇa Maharshi, (Editor), Rider and Company, London, 1988

28. Monod, Jacques., Chance and Necessity, Collins: London, 1972

29. Penfield, Wilder., The Mystery of the Mind: A Critical study of Consciousness and the Human Brain, Princeton University Press, 1975

30. Penrose, Roger., The Emperor's New Mind, Oxford University

31. Penrose, Roger., Shadows Of The Mind, Oxford University Press, NY, 1994 Press, 1989

32. Pope John Paul II., Crossing The Threshold Of Hope, Alfred A. Knopf, 1994

33. Prabhupāda, Swami., Dialectic Spiritualism: A Vedic view of western philosophy, Prabhupada books, New Vrindaban, Moundsville, W.Virginia, 1985

34. Viśiṣṭādvaita philosophy and religion, A collection of essays, Rāmānuja Research Society, T.Nagar, Madras, India, 1974

35. Nikhilānanda, Swami., The Māndukya Upaniṣad (with Śaṁkara's Commentary), (Translator) 1987, Advaita Aśrama, Calcutta.

36. Ranganāthānanda, Swami., The Message of The Upaniṣads, Bhāratīya Vidyā Bhavan, Bombay, 1968

37. Rucker, Rudy., Infinity and the Mind, Rudy Rucker, Bantam Books, 1982

38. Sarasvatī., Śrī Chandrashekhara., The Vedas, Bhāratīya Vidyā Bhavan, 1988,

39. Russell, Bertrand., A History of Western Philosophy, Simon and Schuster, 1945

40. Russell, Bertrand., Why I am not a Christian, Unwin Books, 1967

41. Schrödinger, Edwin, What Is Life?, The Physical Aspect Of The Living Cell, Cambridge University Press, Cambridge, 1944

42. Singh, T.D., Synthesis of Science and Religion, Critical Essays and Dialogues, Editor, The Bhaktivedānta Institute, San Francisco and Bombay, 1987

43. Śrīrangaguru, Amaravāṇi series of books, (In Kannada language), A posthumous collection of lectures, Aṣṭānga Vijñāna Mandiram, Mysore, Karnātaka, India.

44. Subbaramaiya, D.S., Śrī Dakṣiṇāmurtistotram Vols 1 and 2, Dakṣiṇāmaya Śrī Śāradā Pītham, Śringeri, Karnātaka, India, 1988

45. Tipler, Frank., The Physics of Immortality, Double day, 1994

46. Underhill, Evelyn., Mysticism, Doubleday, 1990

47. Wallace R.K., and Benson, H., The Physiology of Meditation, Scientific American 1972

48. Wolters, Clifton., The Cloud of Unknowing and other works, Modern English Translation, Penguin Classics, 1961, 1978

49. Zimmer, Heinrich., Philosophies of India, Edited by Joseph Campbell, Bollingen Series, Princeton, 1969

About The Author

Dr. H.K. Kesavan received his undergraduate degrees in science and engineering from Bangalore, India, and his postgraduate degrees in electrical engineering from the Universities of Illinois and Michigan State in the U.S.A. During his career, he has served as chairman of the department of electrical engineering at the University of Waterloo, Ontario, Canada, and later as the founding chairman of the department of systems design engineering at the same university. In between, for a period of five years, he was the first head of the electrical engineering department at the Indian Institute of Technology, Kanpur, and also, concurrently, the head of its computer center. Later, he was also the first Dean of research and development at the Institute. At present, he is a Distinguished Professor Emeritus at the University of Waterloo.

Dr. Kesavan's research interests encompass two broad fields; first, system theory and secondly, the study of probabilistic systems based on the optimization of information-theoretic principles. He has numerous research publications in both areas and has coauthored several books.

Dr. Kesavan has throughout maintained his professional contacts with the higher Institutes of science and technology in India. He is an honorary fellow of the Computer Society of India.

Dr. Kesavan has maintained a life-long interest in the philosophical aspects of science. This book presents a global view of *science and spirituality* from a Hindu perspective. The insights provided by science for the understanding of Vedic philosophy are highlighted.

Printed in the United States
51222LVS00004B/208-234

9 781410 783776